VCE Units 3 & 4
BIOLOGY

CARISSA KELLY | STACEY MARTIN

A+

2022–2026 STUDY DESIGN • 2022–2026 STUDY DESIGN •

+ course study notes
+ exam practice questions
+ detailed, annotated solutions
+ study and exam preparation advice

STUDY NOTES

A+ Biology Study Notes VCE Units 3 & 4
1st Edition
Carissa Kelly
Stacey Martin
ISBN 9780170479431

Publisher: Alice Wilson
Project editor: Felicity Clissold
Editor: Kelly Robinson
Proofreader: Marcia Bascombe
Cover design: Nikita Bansal
Text design: Alba Design
Project designer: Nikita Bansal
Permissions researcher: Wendy Duncan
Production controllers: Renee Tome, Karen Young
Typeset by: SPi Global
Reviewers: Cathy Jackson, Dawn Duncan

Any URLs contained in this publication were checked for currency during the production process. Note, however, that the publisher cannot vouch for the ongoing currency of URLs.

Acknowledgements
Selected VCE Examination questions and extracts from the VCE Study Designs are copyright Victorian Curriculum and Assessment Authority (VCAA), reproduced by permission. VCE® is a registered trademark of the VCAA. The VCAA does not endorse this product and makes no warranties regarding the correctness or accuracy of this study resource. To the extent permitted by law, the VCAA excludes all liability for any loss or damage suffered or incurred as a result of accessing, using or relying on the content. Current VCE Study Designs, past VCE exams and related content can be accessed directly at www.vcaa.vic.edu.au

For product information and technology assistance,
in Australia call **1300 790 853**;
in New Zealand call **0800 449 725**

For permission to use material from this text or product, please email **aust.permissions@cengage.com**

ISBN 978 0 17 047943 1

Cengage Learning Australia
Level 7, 80 Dorcas Street
South Melbourne, Victoria Australia 3205

Cengage Learning New Zealand
Unit 4B Rosedale Office Park
331 Rosedale Road, Albany, North Shore 0632, NZ

For learning solutions, visit **cengage.com.au**

Printed in China by 1010 Printing International Limited.
3 4 5 6 7 25 24 23

CONTENTS

UNIT 3

HOW DO CELLS MAINTAIN LIFE?

UNIT 4

HOW DOES LIFE CHANGE AND RESPOND TO CHALLENGES?

HOW TO USE THIS BOOK

The A+ Biology resources are designed to be used year-round to prepare you for your VCE Biology exam. *A+ Biology Study Notes* includes topic summaries of all key knowledge in the *VCE Biology Study Design 2022–2026* that you will be assessed on during your exam. Each chapter of this book addresses one Area of Study from the study design. This section gives you a brief overview of each chapter and the features included in this resource.

Area of Study summaries

The Area of Study summaries at the beginning of each chapter give you a high-level summary of the essential knowledge and key science skills you will need to demonstrate during your exam.

Concept maps

The concept maps at the beginning of each chapter provide a visual summary of the key knowledge of each Area of Study outcome. They occur at the beginning of each chapter, after the area of study summary.

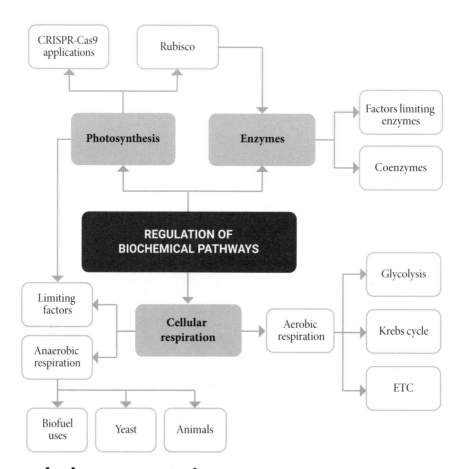

Key knowledge summaries

Key knowledge summaries in each chapter sequentially address **all** key knowledge of the study design.

Revision summary

Use the table to complete brief descriptions of all key knowledge following the key knowledge summaries in each chapter.

Exam practice

Exam practice questions appear at the end of each chapter to test you on what you have just reviewed in the chapter.

Multiple-choice questions

Each chapter has approximately 50 multiple-choice questions.

Short-answer questions

There are 20 short-answer questions in each chapter, often broken into parts. These questions require you to apply your knowledge across multiple concepts. Mark allocations have been provided for each question.

Solutions

Solutions to practice questions are supplied at the back of the book. They have been written to reflect a high-scoring response and include explanations of what makes an effective answer.

Explanations

The solutions section includes explanations of each multiple-choice option, both correct and incorrect, and explanations to written response items explain what a high-scoring response looks like and signposts potential mistakes.

1 A

Due to complementary base pairing, cytosine pairs with guanine; therefore, there will be the same number of each.

B is incorrect because whether there is twice the amount of cytosine as thymine depends on the sequence of bases on the segment of DNA. **C** is incorrect because it depends on the sequence of bases on the segment of DNA as to whether all four are in equal amounts. **D** is incorrect because there is no uracil in DNA.

34 a The ribosome binds to the mRNA and tRNA brings in specific amino acids. (1 mark) The tRNA anticodon is complementary to the mRNA codon. (1 mark) The amino acids are joined by condensation polymerisation to produce the protein enzyme lipase. (1 mark)

In your answer, always remember to refer to the question being asked.

Icons

You will notice the below icons occurring in the summaries and exam practice sections of each chapter.

This icon appears next to official past VCAA questions.

One of these icons appears next to all questions to indicate whether the question is easy, medium or hard.

Hint, reminder and note boxes appear throughout the key knowledge summaries to provide additional tips and support for certain key knowledge.

About *A+ Practice Exams*

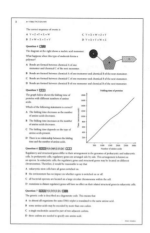

A+ Biology Study Notes can be used independently, or alongside the accompanying resource *A+ Biology Practice Exams*. *A+ Biology Practice Exams* features 14 topic tests comprises official VCAA questions and original VCAA-style questions. Each topic test includes multiple-choice and short-answer questions and focuses on one key knowledge area of the VCAA *VCE Biology Study Design 2022–2026*. There are two complete Units 3 & 4 practice exams after the tests. Like in *A+ Biology Study Notes*, detailed solutions are included at the end of the book, demonstrating and explaining how to craft high-scoring exam responses.

A+ DIGITAL FLASHCARDS

Revise key terms and concepts online with the A+ Flashcards. Each topic glossary in this book has a corresponding deck of digital flashcards you can use to test your understanding and recall. Just scan the QR code or type the URL into your browser to access them.

    Note: You will need to create a free NelsonNet account.

https://get.ga/a-biology-vce-u34

PREPARING FOR THE END OF YEAR EXAM

Exam preparation is a year-long process. It is important to keep on top of the theory and consolidate often, rather than leaving work to the last minute. You should aim to have the theory learnt and your notes complete so that by the time you reach SWOT Vac, the revision you do is structured, efficient and meaningful.

Effective preparation involves the following steps.

Study tips

To stay motivated to study, try to make the experience as comfortable as it can be. Have a dedicated study space that is well lit and quiet. Create and stick to a study timetable, take regular breaks, reward yourself with social outings or treats and use your strengths to your advantage. For example, if you are great at art, turn your Biology notes into cartoons, diagrams or flow charts. If you are better with words or lists, create flash cards or film yourself explaining tricky concepts and watch the videos back.

Revision techniques

Here are some useful revision methods to help information **STIC**.

Spaced repetition	This technique helps to move information from your short-term memory into your long-term memory by spacing out the time between your revision and recall flash card sessions. As the time between retrieving information is slowly extended, the brain processes and stores the information for longer periods.
Testing	Testing is necessary for learning and is a proven method for exam success. If you test yourself continually before you learn all the content, your brain becomes primed to retain the correct answer when you learn it.
Interleaving	This is a revision technique that sounds counterintuitive but is very effective for retaining information. Most students tend to revise a single topic in a session, and then move onto another topic in the next session. With interleaving, you may choose three topics (1, 2, 3) and spend 20 to 30 minutes on each topic. You may choose to study 1-2-3 or 2-1-3 or 3-1-2, 'interleaving' the topics and repeating the study pattern over a long period of time. This strategy is most helpful if the topics are from the same subject and are closely related.
Chunking	An important strategy is breaking down large topics into smaller, more manageable 'chunks' or categories. Essentially, you can think of this as a branching diagram or mind map where the key theory or idea has many branches coming off it that get smaller and smaller. By breaking down the topics into these chunks, you will be able to revise the topic systematically.

These strategies take cognitive effort, but that is what makes them much more effective than re-reading notes or trying to cram information into your short-term memory the night before the exam!

Time management

It is important to manage your time carefully throughout the year. Make sure you are getting enough sleep, that you are getting the right nutrition, and that you are exercising and socialising to maintain a healthy balance so that you don't burn out.

To help you stay on target, plan out a study timetable. One way to do this is to:

1 Assess your current study time and social time. How much are you dedicating to each?
2 List all your commitments and deadlines, including sport, work, assignments, etc.
3 Prioritise the list and re-assess your time to ensure you can meet all your commitments.
4 Decide on a format, whether it be weekly or monthly, and schedule in a study routine.
5 Keep your timetable somewhere you can see it.
6 Be consistent.

Studies suggest that 1-hour blocks with a 10-minute break is most effective for studying, and remember you that can interleave three topics during this time! You will also have free periods during the school day you can use for study, note-taking, assignments, meeting with your teachers and group study sessions. Studying does not have to take hours if it is done effectively. Use your timetable to schedule short study sessions often.

The exam

The examination is held at the end of the year and contributes 50% to your study score. You will have 150 minutes plus 15 minutes of reading time. You are required to attempt multiple-choice questions – Section A, and short answer questions – Section B, covering all areas of study in Units 3 & 4. The following strategies will help you prepare for the exam conditions.

Practise using past papers

To help prepare, download the past papers from the VCAA website and attempt as many as you can in the lead-up to the exam. These will show you the types of questions to expect and give you practice in writing answers. It is a good idea to make the trial exams as much like the real exam as possible (conditions, time constraints, materials etc.).

Use trial papers, School Assessed Coursework, and comments from your teacher to pinpoint weaknesses, and work to improve these areas. Do not just tick or cross your answers; look at the suggested answers and try to work out why your answer was different. What misunderstandings do your answers show? Are there gaps in your knowledge? Read the examiners' reports to find out the common mistakes students make.

Make sure you understand the material, rather than trying to rote learn information. Most questions are aimed at your understanding of concepts and your ability to apply your knowledge to new situations.

The day of the exam

The night before your exam, try to get a good rest and avoid cramming, as this will only increase stress levels. On the day of the exam, arrive at the venue early and bring everything you will need with you. If you have to rush to the exam, your stress levels will increase, thereby lowering your ability to do well. Further, if you are late, you will have less time to complete the exam, which means that you may not be able to answer all the questions or may rush to finish and make careless mistakes. If you are more than 30 minutes late, you may not be allowed to enter the exam. Do not worry too much about exam jitters. A certain amount of stress is required to help you concentrate and achieve an optimum level of performance. If, however, you are feeling very nervous, breathe deeply and slowly. Breathe in for a count of six seconds, and out for six seconds until you begin to feel calm.

Important information from the study design

There are some important terms within the study design that are not explicitly stated in the key knowledge dot points; rather, they can be applied to or taught within the context of the outcomes throughout the year. They are examinable, and you are expected to know them.

Approach to bioethics terms	Definition
Consequences-based	This approach focuses on positive outcomes while minimising negative outcomes.
Duty/rule-based	This approach focuses on processes or rules that must be followed regardless of the consequences or outcomes.
Virtues-based	This approach focuses on the person carrying out the action, and their character or moral values.
Integrity	The honest reporting of all sources of information and communication of results even if they are unfavourable

>>

Approach to bioethics terms	Definition
Justice	The fair consideration of competing claims, and the ideal distribution of risks and benefits so no group is left burdened
Beneficence	Researchers must have the well-being of the research participant as a goal of any experiment or research study.
Non-maleficence	An obligation not to inflict any harm on anyone
Respect	Recognition that a living being is autonomous, unique and free, and has the right and capacity to make their own decisions

Aboriginal and Torres Strait Islander Knowledge terms	Definition
Country	A spiritual term rather than geographical, encompassing an area that a group may look after but that they also have a deep connection to
Place	A geographical area that a group regards as its own; can also have more spiritual connotations

Other definitions include errors, uncertainty and outliers, which are covered in Unit 4, Outcome 3, Area of Study 3.

Details about Section A

Section A consists of a question book and an answer sheet. The answers for multiple-choice questions must be recorded on the answer sheet provided. A correct answer scores 1, and an incorrect answer scores 0. There is no deduction for an incorrect answer, so attempt every question. Read each question carefully and underline key words. If you are given a graph or diagram, make sure you understand the graphic before you read the answer options. You may make notes on the diagrams or graphs.

ONE ANSWER PER LINE

	A	B	C	D
1	A	B	C	D
2	A	B	C	D
3	A	B	C	D
4	A	B	C	D
5	A	B	C	D
6	A	B	C	D
7	A	B	C	D

Details about Section B

Section B consists of a question book with space to write your answers. The space provided is an indication of the detail required in the answer. Each question will be broken down into several parts, and each part will be testing new information; so, read the entire question carefully to ensure you do not repeat yourself. Use correct biological terminology and make an effort to spell it correctly! Look at the mark allocation. Generally, if there are two or three marks allocated to the question, you will be expected to make two or three relevant points. If you make a mistake, cross out any errors but do not write outside the space provided; instead, ask for another booklet and re-write your answer. Mark clearly on your paper which questions you have answered where.

Make sure your handwriting is clear and legible and attempt all questions! Marks are not deducted for incorrect answers, whereas you might get some points if you make an educated guess. You will definitely not get any marks if you leave a question blank!

Do not be put off if you do not recognise an example or context; questions will always be about the concepts that you have covered.

Reading time

Use your time wisely! *Do not* use the reading time to try and figure out the answers to any of the questions until you have read the whole paper! The exam will not ask you a question testing the same knowledge twice, so look for hints in the stem of the question and avoid repeating yourself. Plan your approach so that when you begin writing you know which section, and ideally which question, you are going to start with. You do not have to start with Section A.

Strategies for answering Section A

Read the question carefully and underline any important information to help you break the question down and avoid misreading it. Read all the possible solutions and eliminate any clearly wrong answers. You can annotate or write on any diagrams or infographics and make notes in the margins. Fill in the multiple-choice answer sheet carefully and clearly. Check your answer and move on. Do not leave any answers blank.

Strategies for answering Section B

The examiners' reports always highlight the importance of planning responses before writing. To do this, **BUG** the question:

Box the command word (describe, explain, state).
Underline any key terms or important information, such as the mark allocation.
Go back and check the question again.

Many questions require you to apply your knowledge to unfamiliar situations, so it is okay if you have never heard of the context before. You should, however, know which part of the study design you are being tested on and what the question is asking you to do. Plan your response in a logical sequence. If the question says, 'describe and explain' then structure your answer in that order. You can use dot points to do this, but ensure you write in full sentences. Rote-learned answers are unlikely to receive full marks, so you must relate the concepts of the study design back to the question and ensure that you answer the question that is being asked. Planning your response to include the relevant information and the key terminology will help you avoid writing too much, contradicting yourself, or 'waffling' on and wasting time. If you have time at the end of the paper, go back and re-read your answers.

Good luck. You've got this!

ABOUT THE AUTHORS

Stacey Martin

Stacey Martin began her teaching career with CSIRO Education after developing a love for teaching while working as a naturalist on the Great Barrier Reef. On completing her teaching qualification, she moved to the Scottish Highlands and taught A-Level Biology at Gordonstoun School for five years. Since returning to Melbourne, Stacey has taught at Haileybury College and Korowa Anglican Girls' School. Stacey was a 2020 VCAA VCE Biology assessor.

Carissa Kelly

Before beginning her teaching career, Carissa Kelly worked in a range of pathology departments for 12 years, including microbiology, biochemistry and veterinary medicine. She has worked as the research and development scientist in a molecular biology department and developed new diagnostic tests for human and animal diseases. Since completing her Master of Teaching, Carissa has taught VCE Science at St Paul's Anglican Grammar School where she is currently Head of Science. Carissa is a former VCAA assessor.

UNIT 3
HOW DO CELLS MAINTAIN LIFE?

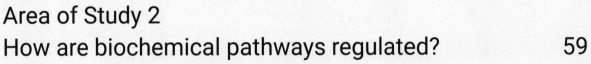

Chapter 1 Area of Study 1
What is the role of nucleic acids and proteins in maintaining life?

Area of Study summary

The focus of Unit 3 is on how cells are able to maintain life, including how the universal code consisting of four nitrogenous bases is stored in genes, and is copied and translated to synthesise proteins. The role of proteins in life-giving processes is also addressed, including enzymes that are integral to the survival of cells in their ability to catalyse the transformation of molecules to provide energy. Gene structure and expression within prokaryotes and eukaryotes is also examined, including the benefits and consequences of their use in manipulating DNA in organisms in the fields of medicine, industry and farming.

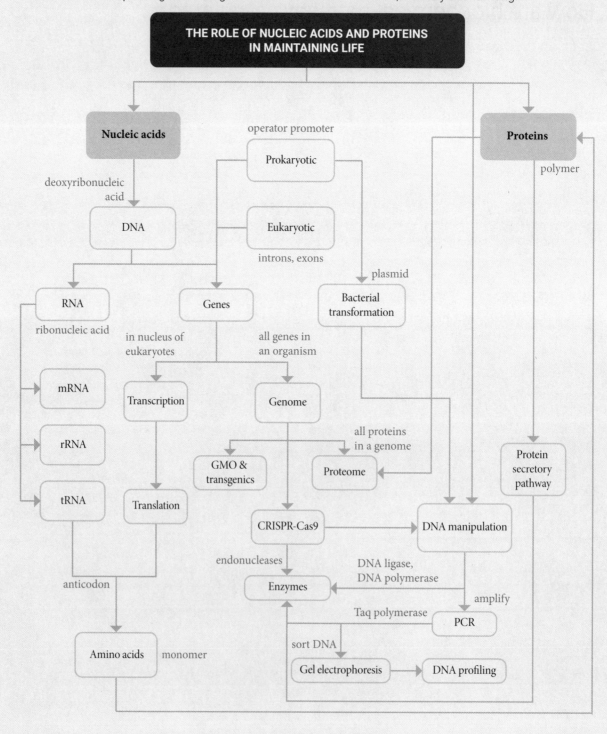

Area of Study 1 Outcome 1

On completing this outcome, you should be able to:

* **analyse** the relationship between nucleic acids and proteins

* **evaluate** how tools and techniques can be used in the manipulation of DNA.

The key science skills demonstrated in this outcome are:

* **analyse**, **evaluate** and **communicate** scientific ideas

* **create** evidence-based arguments and draw conclusions.

© (2021). The Victorian Curriculum and Assessment Authority (VCAA). Used with permission.

Summary of Units 1 & 2 prior knowledge

Cellular structure and function

Cells as the basic structural feature of life

Atoms are the basic unit of matter, including molecules such as proteins, carbohydrates and lipids. These molecules combine to make cells, the smallest life form on Earth.

Prokaryotic cells

Prokaryotic cells are the smallest type of cells. Bacteria, archaebacteria and eubacteria are prokaryotes. Most bacteria are very small, with diameters from 0.5 to 1.0 micrometres (µm) (1 µm = 0.001 mm). Prokaryotes are never truly multicellular, but some form colonies of cells.
Prokaryotic cells:

* have a plasma membrane and cytosol

* lack membrane-bound organelles

* have no nuclear membrane

* have just one circular chromosome.

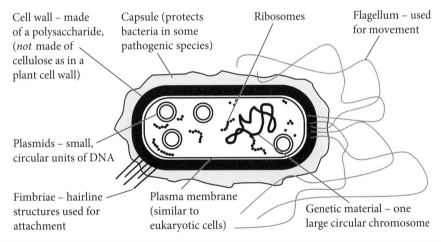

A bacterium as seen with an electron microscope

Eukaryotic cells

Eukaryotic cells are larger than prokaryotic cells, ranging in size from 10 µm to 100 µm. Animals, plants, fungi and protists have eukaryotic cells. (Protists are unicellular or very simple multicellular life forms.)
    Eukaryotic cells are often specialised and organised into complex multicellular life forms They:

* contain membrane-bound **organelles** suspended in a fluid and semi-fluid matrix known as **cytosol** (The membrane, organelles and cytosol make up the **cytoplasm**.)

* have a distinct double-layered nuclear membrane

* have many linear chromosomes.

Limitations of cell size and the need for internal compartments

All cells need to exchange materials with their environments. They must take in gases, water and nutrients, and get rid of waste products. Exchange of materials can only occur through the cell's plasma membrane. The larger a cell becomes, the smaller its surface area becomes in relation to its volume. Look at these cubes to see how increasing size affects the ratio between surface area and volume.

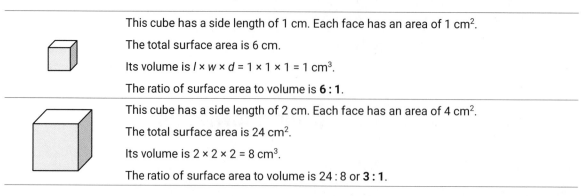

	This cube has a side length of 1 cm. Each face has an area of 1 cm².
	The total surface area is 6 cm.
	Its volume is $l \times w \times d = 1 \times 1 \times 1 = 1$ cm³.
	The ratio of surface area to volume is **6 : 1**.
	This cube has a side length of 2 cm. Each face has an area of 4 cm².
	The total surface area is 24 cm².
	Its volume is $2 \times 2 \times 2 = 8$ cm³.
	The ratio of surface area to volume is 24 : 8 or **3 : 1**.

When a cell grows, its volume increases at a greater rate than its surface area. If the cell becomes too large it is unable to exchange sufficient materials with its environment. Prokaryotic cells, which lack storage organelles and specialised organelles for cellular processes, must be even smaller to allow the exchange of sufficient materials for their life processes.

Specialised cells and organelles

Cells become specialised with internal organelles; structures that are adapted to carry out particular functions. It is important to distinguish which organelles are found in which cell types and to be able to label and identify the structures and locations within the **mitochondria** and the **chloroplasts** that are essential for carrying out chemical processes essential for life.

Plant cell structure

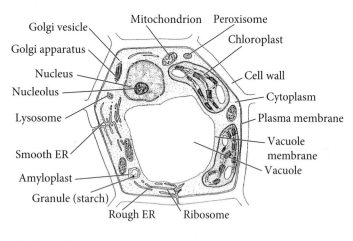

A plant cell

Animal cell structure

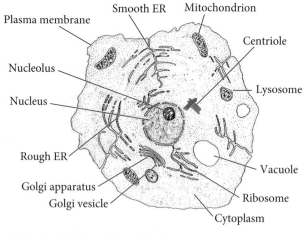

An animal cell

9780170479431

Cell organelles

Name and location	Description	Function	Present in:		
			Prokaryotes	Eukaryotes	
				Animal	Plant
Plasma membrane (also called the cytoplasmic or cell membrane)	A double layer of phospholipids embedded with **protein molecules**, **glycoproteins** and some short carbohydrate chains	Encloses cell contents Regulates the movement of materials into and out of the cell Helps to maintain shape Communicates with neighbouring cells	✓	✓	✓
Nucleus	Surrounded by a double membrane and is usually large compared to other organelles; contains chromosomes and the **nucleolus**	Information in the chromosomes controls the production of proteins in the cell. These proteins, in turn, control cellular functions.	✗	✓	✓
Nuclear membrane	A double membrane, containing many pores, that separates the nucleoplasm from the rest of the cell	Regulates the movement of material between the cytoplasm and the nucleus	✗	✓	✓
Nucleolus (1–3 inside the nucleus)	A granular structure in the nucleus, made of protein and nucleic acid	Site of manufacture of ribosomal RNA	✗	✓	✓
Mitochondria (singular: mitochondrion) Few to thousands in the cytoplasm	Sacs made up of two membranes. The inner membrane is folded to create cristae, which increase the surface area. The cristae enclose the fluid-filled matrix.	The site of aerobic cellular respiration, they are responsible for most of the cell's energy production Most abundant in cells with high energy needs; e.g. muscle cells and cells of the kidney tubule	✗	✓	✓
Ribosomes Some are free in the cytosol, others are attached to rough ER	Small spherical bodies composed of RNA and protein	The site of protein synthesis	✓	✓	✓
Smooth endoplasmic reticulum Found throughout the cytoplasm	A network of membranes that create channels in the cytoplasm from the nucleus to the cell membrane	The site of lipid synthesis and transport of many materials throughout the cell Abundant in cells that produce steroid hormones	✗	✓	✓

»

Name and location	Description	Function	Present in:		
			Prokaryotes	Eukaryotes	
				Animal	Plant
Rough endoplasmic reticulum Found throughout the cytoplasm	A network of membranes that create channels in the cytoplasm Rough ER is studded with ribosomes	Transports and folds proteins produced by ribosomes Synthesises glycoproteins Parts of the ER pinch off to become vesicles for transport of proteins to the membrane or to other organelles, including the Golgi bodies	✗	✓	✓
Golgi apparatus	Stacks of flattened membrane sacs.	Packages proteins and glycoproteins into vesicles for secretion from the cell or to be sent to other organelles Modifies some proteins and glycoproteins Synthesises cellulose in plant cells Abundant in cells that export cell products	✗	✓	✓
Lysosome Found throughout the cytoplasm of animal cells	A membrane sac containing many digestive enzymes	Contains enzymes that digest ingested materials and wastes Important in programmed cell death	✗	✓	✗
Vacuole One large in mature plant cells; one to many small in animal cells	A membrane-bound sac containing fluid, sugars and ions; appears as a clear area when seen through a microscope	Important in turgor in plant cells; the pressure of fluid in the large vacuole helps maintain cell shape Also used for storage of sugars, ions and food	✗	✓	✓
Chloroplast Few to many in cells that photosynthesise Granum Starch grain Stroma (fluid containing enzymes)	Surrounded by a double membrane, contains stacks of membranes (grana) embedded in a less dense membrane and fluid (stroma)	The site of photosynthesis; grana (or thylakoid membranes) contains chlorophyll, the pigment that traps light energy Abundant in leaf cells	✗	✗	✓ some

9780170479431

Name and location	Description	Function	Present in:		
			Prokaryotes	Eukaryotes	
				Animal	Plant
Centriole Near the nucleus in animal cells	A pair of cylindrical structures made up of several smaller tubes	The spindle forms between centrioles during cell division	✗	✓	✗
Flagella or **cilia** Extension(s) of the plasma membrane	Made of membrane material and microtubules, they protrude from the cell Flagellum Cilia Prokaryotic flagella lack membrane material	Provide a means of movement of the cell or of the fluid surrounding the cell	✓ some	✓ some	✓ some
Cell wall *Not* really an organelle; *outside* the cell	A layer of cellulose (in plants) secreted by the cell and completely surrounding the plasma membrane Other chemicals make up the cell walls of bacteria and fungi	Provides structural support and defines the shape of cells	✓	✗	✓

The structure and function of the plasma membrane

All cells are surrounded by a plasma membrane. Plasma membranes are selectively (semi-/partially/ differentially) permeable, allowing some (but not all) substances to pass between the internal and external environments.

Plasma membranes are composed of a **phospholipid bilayer**, with various proteins embedded throughout the phospholipid. The phospholipid and protein components of the membrane enable the passage of different molecules. Carbohydrate chains are involved in cell recognition.

The phospholipid bilayer and its embedded molecules behave like a fluid rather than a solid. These observations are explained using the fluid mosaic model for plasma membranes. The fluidity provides elasticity to the membrane. Proteins, glycoprotein and other molecules move within the membrane, and the lipid bilayer itself can readily change shape.

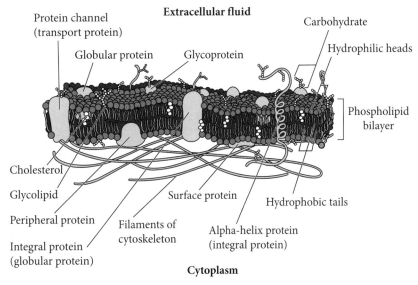

Fluid mosaic model of the plasma membrane

Diffusion

A solution is a mixture of two or more substances. The solvent does the dissolving and the solute is the substance that is dissolved. **Diffusion** is the tendency of particles of gases, liquids and solutes to disperse randomly and fill the available space. The rate of diffusion increases with increased temperature. Small particles in solution can pass through a differentially permeable membrane from an area of high **solute** concentration to an area of low solute concentration. This is often described as moving along a **concentration gradient**. Eventually, a solution will have an even concentration of solute throughout; but this does not mean the particles have stopped moving. The movement is random so there is no *net* movement of particles.

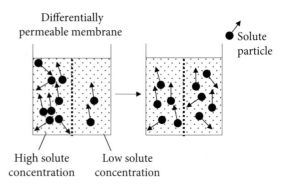

Osmosis

Osmosis is the net movement of water molecules, across a semi-permeable membrane, from an area in which water is in high concentration (low solute concentration) to an area in which water is in low concentration (high solute concentration).

When referring to water, we don't usually use the word 'concentration'. Where there is low solute concentration, water is said to have a high water potential. Water will move from an area of high water potential to an area of lower potential.

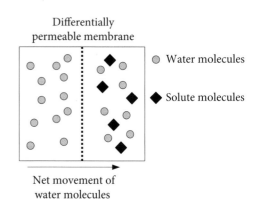

Osmosis and cells

The intercellular fluid surrounding cells is usually **isotonic** to intracellular (within the cell) fluid. This means it has a solute concentration that is the same as that inside the cell. When a cell is in an isotonic solution there is no net movement of water into or out of the cell. If, however, the solute concentration in the fluid surrounding the cell is different from that inside the cell, the cells gain or lose water by osmosis.

Surrounding fluid	Animal cell	Plant cell
Isotonic solution – solute concentration is the same as that inside the cell	Water moves into and out of the cell at the same rate. There is no net movement of water into or out of the cell.	Water moves into and out of the cell at the same rate. There is no net movement of water into or out of the cell.
Hypotonic solution – solute concentration is lower than that inside the cell	Water enters the cell, causing it to swell. It may swell enough to rupture the plasma membrane. This is known as lysis, as 'lysis' means 'to cut apart'.	Water enters the cell and passes into the vacuole. The vacuole swells, squeezing the cytoplasm against the cell wall. The cell becomes turgid.

Surrounding fluid	Animal cell	Plant cell
Hypertonic solution – solute concentration is higher than that inside the cell	There is net movement of water out of the cell, causing it to shrink and possibly die. This process is known as crenulation and the cell is said to be crenated.	Water leaves the cell, causing the plasma membrane to shrink. This is known as **plasmolysis**. The cell wall maintains cell shape but the cells lose turgor, and become flaccid. We see this as plant wilting.

Movement of materials through plasma membranes

Gases, liquids and solutes tend to move from an area in which they are in high concentration to an area of low concentration. When cells take in materials or get rid of wastes by diffusion, no energy is expended by the cell. Sometimes, however, cells require materials that are in higher concentration inside the cell than outside the cell. In order to take in these materials against a concentration gradient, cells must expend energy. This energy is usually in the form of **ATP (adenosine triphosphate)**.

Simple diffusion	Small uncharged particles and lipid-soluble molecules diffuse from an area of high concentration to an area of low concentration through the phospholipid bilayer. These include O_2, CO_2, H_2O, alcohol and urea. In **simple diffusion**, no energy is used by the cell.	High solute concentration Low solute concentration
Facilitated diffusion	The **facilitated diffusion** of other small, polar molecules from an area of high concentration to an area of low concentration occurs through channels provided by the embedded protein molecules. Transported substances include ions, glucose and amino acids. No energy is used by the cell. **Hint** The phospholipids allow small, non-polar molecules such as gases (carbon dioxide and oxygen) to diffuse through the cell surface membrane. The protein channels allow for large polar molecules (such as glucose) to enter the cell through facilitated diffusion.	High glucose concentration Low glucose concentration (inside cell)
Active transport	**Active transport** is the transport of molecules across the plasma membrane against a concentration gradient. Active transport occurs through the **protein channels** and requires the cell to expend energy in the form of ATP. Plant roots actively transport minerals from the soil into the root hair cells.	Low concentration of potassium High concentration of potassium ATP $ADP + P_i$

›>

››	**Endocytosis** Pinocytosis (fluid) and phagocytosis (large particles and debris)	Cells are able to take in larger quantities of solids and liquids by **endocytosis**. Parts of the plasma membrane fold around the material and pinch off, enclosing the material in a vesicle inside the cell. This process requires energy. 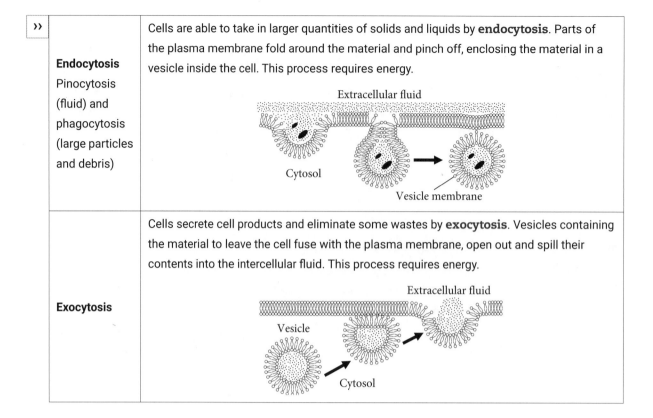
	Exocytosis	Cells secrete cell products and eliminate some wastes by **exocytosis**. Vesicles containing the material to leave the cell fuse with the plasma membrane, open out and spill their contents into the intercellular fluid. This process requires energy.

1.1 The relationship between nucleic acids and proteins

Nucleic acids are comprised of **deoxyribonucleic acid (DNA)** and various types of **ribonucleic acid (RNA)**. DNA and RNA are large organic polymers that consist of repeating monomers known as nucleotides. The cell uses the DNA as a set of instructions to build a protein. The DNA (the instructions) is written in a code, and so it must be copied and then translated by the RNA in order to carry out the instructions it carries.

1.1.1 Nucleic acids as information molecules

The structure of nucleic acids

Nucleic acids are biological molecules made up of repeating nucleotides. Each nucleotide consists of a phosphate bonded to a 5-carbon sugar called deoxyribose (hence the name 'deoxyribonucleic acid').

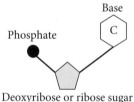

This sugar is attached to a nitrogen-containing base. There are four bases in a DNA molecule: guanine (G), adenine (A), cytosine (C) and thymine (T).

The arrangement of a DNA molecule is such that the nucleotides form **covalent bonds**, called phosphodiester bonds, between the sugar and the phosphate to build a stable molecule. You will notice in the diagram over the page that the sugar–phosphate backbone runs along both sides, with the bases in the middle like a ladder.

The left side of the ladder runs with the 5-carbon facing up, and the right side of the ladder with the 3-carbon facing up. The left side runs from the direction of 5′ to 3′ and the right side is in the direction from 3′ to 5′. (5′ is pronounced '5 prime' and 3′ is pronounced '3 prime'). The strands run antiparallel, or opposite to one another. In the next diagram you can see how the carbons in the pentose sugars, deoxyribose and ribose are arranged. These sugars are referred to as pentose because they have 5 carbons, and *pent(a)-* is the suffix for *5*.

In the 'rungs' of the ladder we have the bases, which pair up according to their chemical properties. Adenine and guanine are both purines. Cytosine and thymine are both pyrimidines. Due to their chemical nature, the base pairings are specific and complementary. That is, A always pairs with T and C always pairs with G. The bases are held together with **hydrogen bonds**. Two hydrogen bonds form between A and T and three hydrogen bonds form between C and G. Complementary base pairing enables DNA to replicate and to transfer information.

The DNA molecule is a called a double helix because it contains the two strands arranged in a helical shape. It is also supercoiled (tightly wound) to allow approximately 3 billion base pairs in each cell to fit into just 6 microns across. If you stretched out the DNA from one of your cells it would be about two metres long.

RNA, like DNA, is made up of repeating nucleotide bases. However, there are some very important differences between DNA and RNA. First, the sugar in RNA is ribose. The RNA molecule is single stranded (not double-stranded like DNA) and it contains four bases: adenine (A), cytosine (C), guanine (G) and uracil (U). Uracil replaces thymine in RNA. Uracil and thymine are chemically similar (they are both pyrimidines), and two hydrogen bonds can form between A and U.

The clockwise numbering of the five carbons in the pentose sugars found in nucleic acids

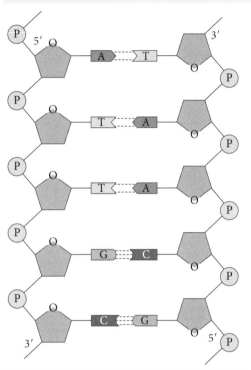

The arrangement of DNA, including the hydrogen bonds between the bases and the anti-parallel sugar–phosphate backbone. Cytosine pairs with guanine (3 hydrogen bonds) and thymine/uracil pairs with adenine (2 hydrogen bonds).

Thymine　　　　Uracil

The similar chemical compositions of thymine and uracil. Uracil replaces thymine in RNA.

9780170479431

Ribonucleic acids

There are three types of RNA, including **messenger RNA (mRNA)**, **ribosomal RNA (rRNA)** and **transfer RNA (tRNA)**. Their roles and their unique appearances are outlined below.

RNA	Role	Diagram
mRNA	mRNA carries the complementary genetic code copied from DNA during transcription to the ribosomes for protein synthesis.	Uracil
rRNA	rRNAs combine with proteins and enzymes in the cytoplasm to form ribosomes. Ribosomes are where protein synthesis occurs.	Ribosome Ribosomal RNA
tRNA	The main function of tRNA is to deliver specific amino acids during protein synthesis. Each of the 20 amino acids has a specific tRNA molecule that binds to it and transfers it to a growing polypeptide chain.	Amino acid acceptor site Anticodon

Comparing RNA to DNA

The following table outlines the differences between DNA and RNA.

RNA	DNA
Single stranded	Double stranded
Ribose sugar	Deoxyribose sugar
ACUG bases	ACTG bases
Short-lived	Long-lasting
Formed 5′ to 3′	Read 3′ to 5′

1.1.2 The genetic code and protein synthesis

DNA is a double helix and is protected in the nucleus, and it is too big to fit through the nuclear pore. However, it contains the code for a protein that needs to be synthesised *outside* the nucleus, at the ribosome. In order for this to happen, DNA is transcribed into mRNA, which is a short-lived copy of the code.

Transcription

The process of transcription is to copy the DNA. The steps involved are outlined below.

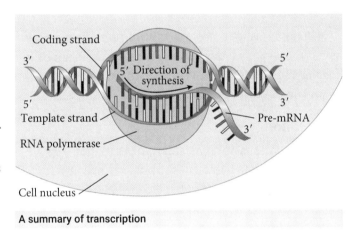

A summary of transcription

1   DNA unwinds temporarily to expose the template strand.

2   RNA polymerase binds to a **promoter** sequence at the start of the gene that is to be copied. RNA polymerase adds free RNA nucleotides to the template strand of DNA. The RNA nucleotides bond together in a condensation polymerisation reaction.

3   Transcription ends by process of termination, which depends on the sequences of mRNA.

4   A pre-mRNA molecule is formed.

RNA processing

Pre-mRNA is the primary product of transcription and must be processed before it becomes mRNA.

In this process, the **introns**, which are non-coding regions, are spliced out by enzymes known as *spliceosomes*.

> **Hint**
> It helps to think of introns as interruptions to the code, which must be removed.

Once the introns are spliced, the **exons** bond together. The exon sections can be arranged in different ways, which would provide a different coding sequence resulting in a different protein. This is known as alternative splicing.

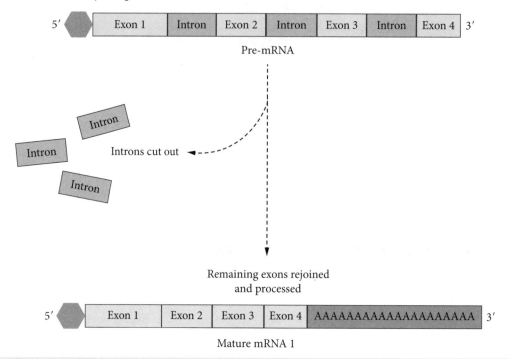

A methyl-cap is added to the 5′ end. The cap is a modified guanine nucleotide known as 7-methylguanosine cap.

Hint
It is fine to refer to this as a *methyl cap*.

Its purpose is to protect the mRNA molecule from degrading, as well as to assist the ribosome binding during translation. Finally, a poly-A tail is added to the 3′ end. This is the addition of many adenine nucleotides, which helps to increase the stability of the molecule, preventing enzymes from degrading the mRNA strand.

Translation

After RNA processing, translation occurs at the ribosome. The mRNA attaches to a ribosome in the cytosol and the code is read three bases at a time. On the DNA, the three bases are referred to as a **triplet**, the corresponding bases on mRNA are known as **codons**, and the complementary bases on tRNA are known as **anticodons**. The mRNA will begin with a start codon (AUG). As the codon passes through the ribosome, the tRNA with the complementary anticodon will have a specific amino acid attached. There are 20 **amino acids** and 64 possible codons. As such, different codons can code for the same amino acid. This is why the DNA code is said to be **degenerate**.

As each amino acid is delivered by the tRNA, it will bond with other amino acids, forming **peptide bonds** between them via **condensation polymerisation reaction**. These are strong covalent bonds that bind them together, forming a polypeptide chain. The tRNA molecule

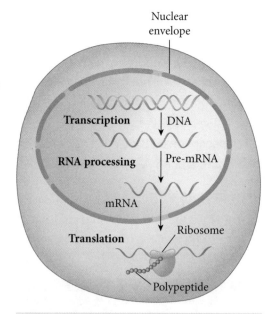

Stages and locations of the steps of proteins synthesis within a eukaryotic cell

will detach from the mRNA and the tRNA can be reused. This process will continue until a stop codon is reached. The mRNA strand can then be used again to produce more of the same protein.

Second base

First base		U	C	A	G	Third base
	U	UUU ⎤ Phe UUC ⎦ UUA ⎤ Leu UUG ⎦	UCU ⎤ UCC UCA ⎬ Ser UCG ⎦	UAU ⎤ Tyr UAC ⎦ UAA　Stop UAG　Stop	UGU ⎤ Cys UGC ⎦ UGA　Stop UGG　Trp	U C A G
	C	CUU ⎤ CUC CUA ⎬ Leu CUG ⎦	CCU ⎤ CCC CCA ⎬ Pro CCG ⎦	CAU ⎤ His CAC ⎦ CAA ⎤ Gln CAG ⎦	CGU ⎤ CGC CGA ⎬ Arg CGG ⎦	U C A G
	A	AUU ⎤ AUC ⎬ Ile AUA AUG　Met/ 　　　Start	ACU ⎤ ACC ACA ⎬ Thr ACG ⎦	AAU ⎤ Asn AAC ⎦ AAA ⎤ Lys AAG ⎦	AGU ⎤ Ser AGC ⎦ AGA ⎤ Arg AGG ⎦	U C A G
	G	GUU ⎤ GUC GUA ⎬ Val GUG ⎦	GCU ⎤ GCC GCA ⎬ Ala GCG ⎦	GAU ⎤ Asp GAC ⎦ GAA ⎤ Glu GAG ⎦	GGU ⎤ GGC GGA ⎬ Gly GGG ⎦	U C A G

Ala = alanine
Arg = arginine
Asn = asparagine
Asp = aspartic acid
Cys = cysteine
Gln = glutamine
Glu = glutamic acid
Gly = glycine
His = histidine
Ile = isoleucine
Leu = leucine
Lys = lysine
Met = methionine
Phe = phenylalanine
Pro = proline
Ser = serine
Thr = threonine
Trp = tryptophan
Tyr = tyrosine
Val = valine

An anticodon table. Notice that there is sometimes more than one anticodon for each amino acid; this is why we call DNA 'degenerate'. However, it is **unambiguous** because each codon is specific to only one amino acid.

1.1.3 The structure of genes

A **gene** is the basic unit of heredity. It is a section on a chromosome that controls the production of one biological molecule, usually a protein. Some genes code for the production of transfer RNA (tRNA) or ribosomal RNA (rRNA). There are about 30 000 genes in the human genome. Each chromosome carries many hundreds of genes. The position of a gene on a chromosome is known as the gene locus (plural, loci).

Promoter regions

Genes contain coding and non-coding sequences of bases. The upstream region of a gene contains base sequences that regulate its activity. It contains a promoter sequence that allows for the recognition and binding of RNA polymerase, which is necessary for transcription.

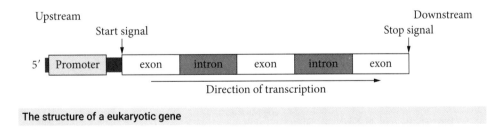

The structure of a eukaryotic gene

Introns and exons

The genes of eukaryotes contain introns and exons. The exons contain the base sequences that code for the amino acids needed to build a protein molecule. The introns are non-coding segments of

DNA that, as discussed, are spliced out during post-transcriptional modification in the nucleus prior to translation. Once the introns are spliced out, the exons bond together and can do so in different arrangements, which is important to protein diversity and **gene expression**. Most eukaryotic genes have several introns.

Operator regions

In prokaryotes, the gene structure differs in that it does not have introns and exons but it does have regions that control gene regulation. The **operator** region of the gene follows the promoter region. The promoter region allows RNA polymerase to recognise the region and bind so the process of transcription can occur. The operator region is where a repressor protein can

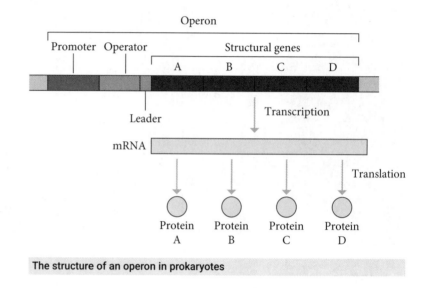

The structure of an operon in prokaryotes

bind. If a repressor is bound to the operator, RNA polymerase is blocked and transcription cannot occur. This allows for organisms to regulate, or switch genes on or off.

1.1.4 Gene regulation

Gene regulation is important in organisms because it enables genes to be turned on or off at particular times, allowing prokaryotes to react to a change in environment. For example, the bacterium *E. coli* must contend with variations in the availability of tryptophan in the environment. Tryptophan is an amino acid essential to building proteins required by the bacterium. If tryptophan is available in the environment, *E. coli* will take it up. If it is not available, *E. coli* will have to synthesise the amino acid through transcription and translation.

Prokaryotic DNA is organised in blocks called operons. Each operon contains genes needed for a specific function, or biochemical pathway. *E. coli* has an operon for synthesising tryptophan, or *trp*, which includes five structural genes side by side. This allows all five required genes to be transcribed simultaneously, and translated into the five enzymes required for tryptophan production.

There are two mechanisms that regulate the expression of the *trp* operon: **repression** and **attenuation**. If tryptophan is available in the environment, the *trp*

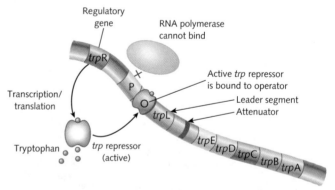

The *trp* operon is 'switched off' when tryptophan is present in the environment.

operon is switched off because *E. coli* do not need to produce it themselves. In order to switch it off, tryptophan molecules bind to a repressor protein, (the repressor protein is coded for by the regulatory gene *trp*R), changing its shape so that it can now bind to the operator sequence on the operon.

If the repressor is bound to the operator, RNA polymerase, which is bound to the promotor, is physically blocked from transcribing the tryptophan genes.

When tryptophan is absent, the repressor protein does not bind to the operator and RNA polymerase is able to transcribe the genes required for synthesising tryptophan.

The second mechanism, attenuation, regulates gene expression without the repressor protein binding to the operator region. Attenuation occurs because of the structure of the leader region of the *trp* operon. The leader region contains 4 sections (1, 2, 3 & 4). Section 1 contains two adjacent *trp* codons (ACC) and in between sections 1 and 2 there is a STOP codon (ACU). Due to the bases that occur in sections 1, 2, 3 and 4, the sections can stick together due to complementary base pairing. Section 1 and 2 can stick together and form a hairpin loop, as can sections 2 and 3, and 3 and 4. At the end of these 4 sections is a string of adenine bases called the attenuator region.

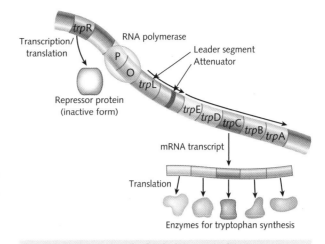

The *trp* operon is 'switched on' when tryptophan is not present in the environment.

As transcription of the DNA by RNA polymerase producing mRNA is occurring the ribosome can immediately start translation. This is because a prokaryotic cell does not have a nucleus and the mRNA and ribosomes occupy the same space. As the ribosome moves along the newly transcribed mRNA it passes by the two *trp* codons (UGG) in section 1. If there is a large or only a small amount of tryptophan in the cell, this will be added to the growing polypeptide chain that is being produced. The ribosome continues on and arrives at the UGA or STOP codon. The ribosome stops overlapping both sections 1 and 2. This means sections 1 and 2; and 2 and 3 are prevented from forming hairpin loops, but sections 3 and 4 stick together and form an attenuator stem loop.

Remember that transcription is still occurring. Also remember the string of adenine bases at the end of the leader section. As the attenuator stem loop is formed between sections 3 and 4 it creates a pulling force on the mRNA. As adenine and uracil are only weakly bonded by two hydrogen bonds the uracil in the mRNA pulls away from the adenine on the DNA and a shortened piece of attenuated mRNA falls away and the ribosome detaches. Transcription has stopped before the RNA polymerase reaches the structural genes.

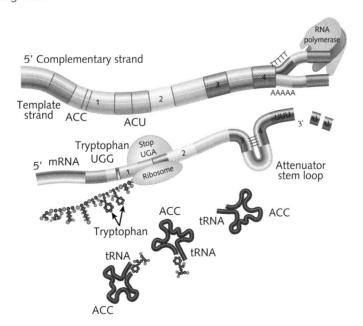

If there is no tryptophan present in the cell the ribosome pauses at the *trp* codons in section 1 and waits for a delivery of tryptophan by a tRNA. This enables sections 2 and 3 to form an anti-attenuator stem loop. This loop formation does not provide enough force to pull the mRNA away from the DNA so transcription of the structural genes continues. Another ribosome will come along and attach to the transcribed mRNA to translate the five structural genes and more tryptophan will be produced.

1.1.5 Amino acids

Amino acids are the building blocks, or monomers, of proteins. A protein is a polymer because it is made of repeating units of amino acids.

There are 20 amino acids that all contain the same basic structure. What differs between the amino acids is the 'R group', in which there

An amino acid group (NH_2)   A carboxyl group

The R group will change depending on the amino acid.

The basic structure of amino acids

are 20 different possibilities. The properties of the R groups in different amino acids are important in determining protein structure, but you do not need to know these for the Biology exam. What is important is that you understand the basic structure, how they bond together to form proteins, and how the initial sequence of amino acids will dictate the structure and function of the protein.

Primary structure

The **primary structure** of a protein is its **amino acid sequence**. The sequence of amino acids is unique for each protein. At the ribosomes, amino acids are assembled into chains called peptides (short chains) or polypeptides (long chains). The fact that there are so many amino acids, each capable of bonding with any other amino acid, means there are millions of possible amino acid sequences.

Bonds between amino acids are formed by a condensation polymerisation reaction (water is lost). Covalent bonds form between a carbon atom of one amino acid and the nitrogen atom of the next. These carbon–nitrogen links are called peptide bonds and are very strong.

The formation of a peptide bond through a condensation reaction between two amino acids

Peptide bonds always occur between the constant parts of the amino acid molecules and never involve the R groups. This strong carbon–nitrogen chain forms the backbone of the protein molecule.

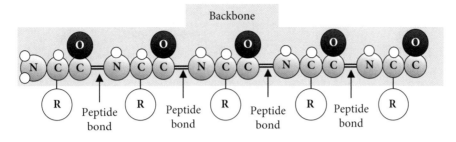

A polypeptide chain polymer forms as many amino acid monomers bond together.

Secondary structure

Long sequences of amino acids do not simply form long straight chains. Oxygen atoms (which have a partial negative charge) and hydrogen atoms (which have a partial positive charge) in the backbone of the polypeptide chain interact with each other, forming hydrogen bonds. These hydrogen bonds cause the peptide chain to coil or pleat, forming a protein's **secondary structure**. There are two common forms of secondary structure: the **alpha helix (α-helix)** and the **beta pleated (β-pleated) sheet**.

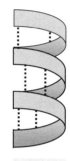

An α-helix

In alpha helices the peptide chain is coiled. The loops of the helix are formed when hydrogen atoms from one part of the molecule's backbone form hydrogen bonds with oxygen atoms from another part of the chain. (Hydrogen bonds are shown as dotted lines.) Proteins that consist largely of alpha helices tend to be elastic (stretchy). They include hair and wool. Wool and hair will stretch, especially when wet, but regain shape as they dry and hydrogen bonds reform.

Beta pleated sheets are also formed by hydrogen bonding between hydrogen and oxygen atoms. Parts of the peptide chain fold back to align with neighbouring parts of the chain. The hydrogen bonding forces the chains into a pleated shape. Unlike alpha helices, beta pleated sheets are not elastic. Silk protein is largely composed of beta pleated sheets. (It doesn't stretch or sag.)

A β-pleated sheet

A protein showing both α-helices and β-pleated sheets

It is not uncommon for a protein to have alpha helices in some regions of its length and beta pleated sheets in other regions.

Tertiary structure

The tertiary structure of a protein is its overall three-dimensional shape and its functional state. Peptide chains, which may have α-helices or β-pleated sheets along their length, fold up further due to interactions between the atoms in their R groups. The bending and folding takes place in the lumen of the rough endoplasmic reticulum and is aided by the action of enzymes known as **chaperones**. This overall shape is determined by:

- hydrogen bonding between atoms in some R groups, and between atoms of the **hydrophilic** amino acids and the surrounding water
- **ionic bonding** between atoms in R groups that are electrically charged
- covalent bonding between the **sulfur** atoms of some R groups (disulfide bridge)
- **hydrophobic interactions**, in which R groups that are **hydrophobic** are repelled by the water-surrounding proteins in the cell, and so tend to collect on the inside of the protein molecules.

> **Hint**
> To help you recall: they get *HICH*ed.

Enzymes exhibit tertiary structure giving them a specific three dimensional shape including their active sites. Some specific enzymes do exhibit quaternary structure.

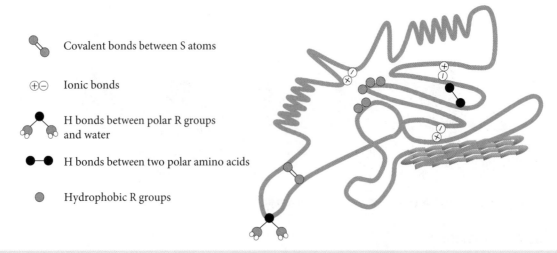

Covalent bonds between S atoms

⊕⊖   Ionic bonds

H bonds between polar R groups and water

H bonds between two polar amino acids

Hydrophobic R groups

The 3D folding of a polypeptide chain is the result of interactions between the R groups and their proximity to one another.

Quaternary structure

Some biologically active proteins consist of more than one polypeptide chain (each with its own primary, secondary and tertiary structures), held together by interactions between atoms in the separate polypeptide chains. Haemoglobin (the oxygen-carrying molecule of red blood cells) consists of four polypeptides. Antibodies (see Chapter 3) consist of four polypeptides held together by disulfide bonds between the polypeptide chains. Some enzymes such as Rubisco, DNA and RNA polymerases exhibit quaternary structure.

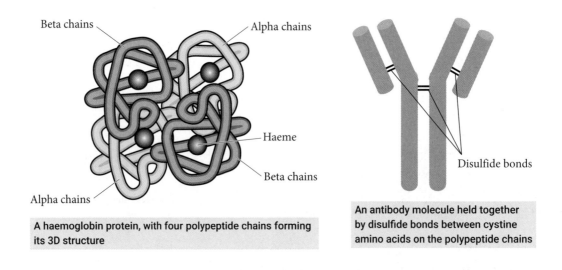

A haemoglobin protein, with four polypeptide chains forming its 3D structure

An antibody molecule held together by disulfide bonds between cystine amino acids on the polypeptide chains

All the 'instructions' for a completed functioning protein are contained within its primary structure. A single gene codes for the production of a functioning protein by dictating the amino acid sequence in that protein (its primary structure). This primary structure determines where, along the protein's length, a chain of amino acids will form α-helices or β-pleated sheets (its secondary structure).

The different R groups present as a result of the primary structure determine the folding and coiling of the tertiary structure and the bonding of peptides to produce the 3D functional structure. The function of a protein is dependent on the shape of the protein and so a different primary structure will result in a different tertiary structure and, hence, the protein's function.

9780170479431

1.1.6 Proteins and an organism's proteome

Proteins are large complex molecules that play a critical role in the structure, function and regulation of an organism. Proteins can be divided into two major groups.

Fibrous proteins are important in **cellular structure**, and include muscle proteins that can contract, proteins that strengthen tendons and ligaments, keratin in nails and skin, and proteins that aid in blood clotting.

Globular proteins include all enzymes, antibodies, many hormones and the proteins that make up cilia and flagella. Proteins in cell membranes act as channels for the movement of substances in and out of cells, and as receptor sites for hormones and other cellular messengers.

All cells contain a multitude of proteins, each with a structure that uniquely fits it for a particular cell function. The sum total of all proteins found in a functioning organism is known as the organism's **proteome**. This will be unique to an individual because the proteins are coded for by an individual's DNA sequence within their genes. The proteome is significantly larger than the genome because alternative splicing of the introns and bonding of the exons ensures further variations of a single gene. Humans have more than 200 000 different proteins. The study of these proteins is known as **proteomics**.

Protein types and their functions

Protein	Role	Examples
Immunoglobulins	Specialised proteins involved in defending the body from foreign invaders	Antibodies circulate in the body and bind with foreign antigens. T cell receptors detect foreign antigens.
Contractile proteins	Involved in cell movement, especially muscle contraction	Actin and myosin contract in muscle movement and in the movement of cells such as phagocytes.
Enzymes	Proteins that catalyse **biochemical reactions**	Lactase breaks down the sugar lactose, found in milk. Pepsin breaks down food proteins.
Hormonal proteins	Some hormones are chemical **messenger proteins**, which play a role in coordination of cellular processes	Insulin regulates glucose metabolism by controlling the blood sugar concentration. Growth hormone stimulates protein production in muscle cells.
Structural proteins	Fibrous and stringy proteins that provide support and strength	Keratins strengthen protective coverings such as hair, feathers and beaks. Collagens and elastin provide support in connective tissue such as ligaments.
Storage proteins	Store amino acids, often for the growth of new organisms, and/or store reserves of mineral ions, such as iron	Ovalbumin in egg white Casein in milk Plant storage proteins in seeds, especially legumes
Transport proteins	Carrier proteins that move molecules from one place to another around the body or within cells	Haemoglobin in red blood cells carries oxygen. Cytochromes are electron carriers in cellular respiration.

Enzymes

Enzymes are a subunit of proteins that rely on optimum conditions in order to lower the activation energy required to carry out metabolic reactions and control biochemical pathways. Enzymes are known as biological catalysts because they increase the rate of reaction without being used in the reaction themselves.

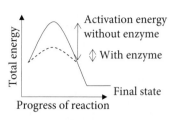

In order to do this, an enzyme will have a unique binding site known as the **active site**. This is complementary and highly specific to the substrate. An enzyme will bind to a specific substrate like a lock and key. For example, the enzyme lactase will bind to the substrate lactose and form an enzyme–substrate complex. The complex facilitates the breaking of bonds in the substrate, causing lactose to be broken down into its two smaller subunits: glucose and galactose. The enzyme does not take part in the reaction so the enzyme is able to be re-used to break down the remaining substrate into product.

Enzymes are biological catalysts because they lower the activation energy required to start a reaction.

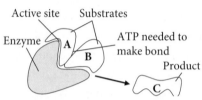

An anabolic, endothermic reaction of lock-and-key process of enzyme action

Reactions in cells

Cells are very active. Chemical reactions are occurring constantly in the cytosol and in cell organelles. Reactions that occur in cells can be divided into two groups: **anabolic reactions** and **catabolic reactions**.

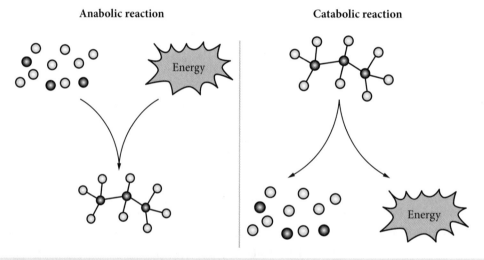

Catabolic reactions vs anabolic reactions

In an anabolic reaction, energy is taken in during the reaction – the cell must expend energy for the reaction to proceed. Examples include the building of complex molecules such as proteins, the manufacture of carbohydrates in photosynthesis, and the synthesis of ATP.

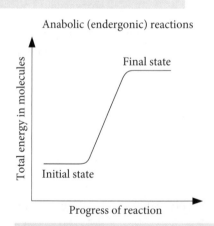

An endothermic reaction where the products contain more energy than the reactants, thus requiring energy from the environment to occur.

9780170479431

In a catabolic reaction, there is a **net gain of energy** from the reaction. Even though the reaction will ultimately produce energy, most exergonic reactions in cells are not spontaneous. They require some expenditure of energy by the cell to get the reaction going (activation energy). As when trying to light a campfire, you put in energy to get it going, and get more energy back. Examples include cellular respiration, and the breakdown of ATP to ADP and P_i.

Catabolic (exergonic) reactions

An exothermic reaction where the products have less energy than the reactants, giving energy off to the environment, usually as heat.

1.1.7 The protein secretory pathway

You now understand how proteins are manufactured at the ribosome, and some of the functions they carry out within the body. In order to get from a ribosome and out of the cell to perform these roles, a secretory pathway is required. This pathway will allow the protein to be moved to the plasma membrane where it can be released in a process called exocytosis.

It is interesting to note the secretory pathway has its own chemical environment separate to the cytosol of the cell. These conditions are unique to the lumen (the liquid that fills the secretory pathway), and the cytosol and lumen do not mix. It is this environment that allows the R groups of the amino acids to fold into their unique 3D structure and for any final modifications of the protein.

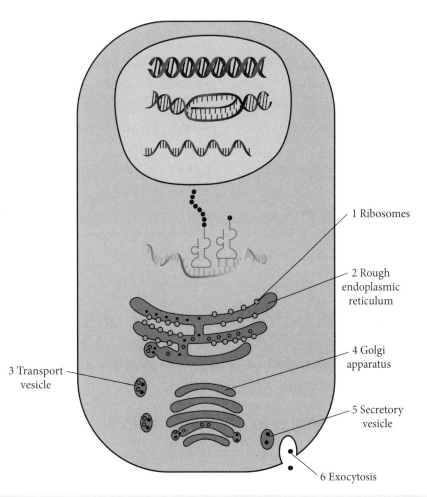

1 Ribosomes

2 Rough endoplasmic reticulum

3 Transport vesicle

4 Golgi apparatus

5 Secretory vesicle

6 Exocytosis

An overview of the protein secretory process and the organelles involved

Step	Location	Role	Explanation
1	Ribosomes	The site of protein synthesis	Ribosomes are attached to the rough endoplasmic reticulum; they translate mRNA into a polypeptide chain.
2	Endoplasmic reticulum	The site of polypeptide folding into a protein	The polypeptide chain enters the continuous tubules of the endoplasmic reticulum, called cisternae, which have a unique chemical environment that allows the polypeptide chain to fold into its unique 3D structure. Some protein modification occurs in the endoplasmic reticulum, such as glycosylation – the addition of a carbohydrate to the protein by enzymes.
3	Transport vesicle	Moves protein toward the *cis* face of the Golgi apparatus	The protein is now ready to leave the endoplasmic reticulum and travel to the Golgi apparatus. In order to protect it from the chemical environment of the cytosol, it travels in a vesicle that buds off from the membrane of the endoplasmic reticulum.
4	Golgi apparatus	The protein is modified and packaged into vesicles	The Golgi apparatus is a set of compartments called sacs or cisternae. Different compartments have different enzymes, which allows the proteins to be modified in order from *cis* to *trans* face and be packaged into a secretory vesicle.
5	Secretory vesicle	Secretory vesicles will fuse with plasma membrane	These vesicles can hang around in the cell until needed. Once needed, they will fuse with the plasma membrane and the proteins will be secreted out of the cell.
6	Exocytosis	Contents of the secretory vesicle are released from the cell using ATP provided by mitochondria	Exocytosis is an energy dependent process. Energy is provided by mitochondria in the form of ATP. Exocytosis moves proteins from inside the cell to the extracellular space; e.g. hormones may travel in the blood where they can bind to receptors on another cell, eliciting a response.
7	Lysosomes	Breaking down misfolded proteins	If there is misfolding within the protein, the secretory vesicle will instead be sent to the lysosome for degradation.

1.2 DNA manipulation techniques and applications

Knowledge of the structure of DNA and the processes of transcription and translation has led scientists to make advancements using technology to manipulate DNA in order to modify genes. This has many applications in medicine, farming and industry.

1.2.1 The use of enzymes to manipulate DNA

Prokaryotes contain enzymes known as **endonucleases**. As the name suggests, they cut within (endo) a nucleotide sequence of DNA. This allows bacteria to cut viral DNA and protect itself from becoming infected.

Endonucleases

Endonucleases that cut DNA at specific recognition sites are known as restriction endonucleases or **restriction enzymes**. They are named after the bacteria from which they were isolated, and each one has a unique recognition site, so it will only cut at a region with the specific base sequence. The table below shows some common restriction enzymes.

Name of enzyme	Source	Recognition site and cleavage site	Nature of cut ends
EcoRI	E. coli RY13	5′ - G|AATTC - 3′ 3′ - CTTAA|G - 5′	Sticky
HindIII	Haemophilus influenzae Rd	5′ - A|AGCTT - 3′ 3′ - TTCGA|A - 5′	Sticky
BamHI	Bacillus amyloliquefaciens H	5′ - G|GATCC - 3′ 3′ - CCTAG|G - 5′	Sticky
BalI	Brevibacterium albidum	5′ - TGG|CCA - 3′ 3′ - ACC|GGT - 5′	Blunt
HaeIII	Haemophilus aegyptius	5′ - GG|CC - 3′ 3′ - CC|GG - 5′	Blunt

Sticky ends and blunt ends

As you can see in the table, the restriction enzyme recognises a sequence and makes a cut along the DNA as if it were a pair of scissors. Restriction enzymes catalyse hydrolysis of the phosphodiester bond along the sugar–phosphate backbone. This means the enzyme speeds up the breaking of the bond holding the sugar–phosphate backbone together at a specific section. This cutting effect can produce **blunt ends**, where there are no nucleotide overhangs, or **sticky ends**, where the cut doesn't run straight through the double strand of DNA and so produces overhanging nucleotides either side of the cut.

Sticky ends are often preferred in DNA manipulation because the overlaps of nucleotides allow them to base pair and join together with another piece of DNA that is complementary. This helps to ensure the DNA is inserted in the right direction.

DNA ligase

In order for the two pieces of DNA from the different species to bond correctly, a phosphodiester bond needs to be formed, gluing the sections together. Another enzyme, **DNA ligase**, catalyses this bond in a condensation reaction between the sugar–phosphate backbone, ligating the two pieces together.

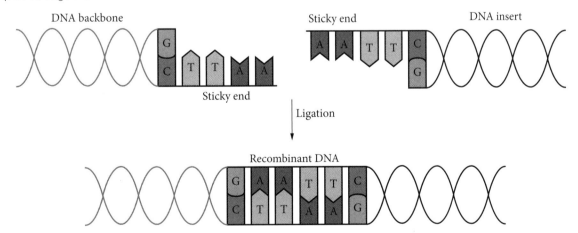

DNA polymerase

DNA polymerase is another enzyme used in DNA manipulation. As its name suggests, it forms a DNA polymer by bonding individual nucleotides, or monomers, together. Its applications are discussed in more detail in section 1.2.3 *Amplification and sorting of DNA fragments*.

Reverse transcriptase

Reverse transcriptase is a polymerase enzyme that transcribes single-stranded RNA into cDNA, or complementary DNA. This can be useful in gene transfer, because the DNA transcribed will not contain introns. This is useful because the mRNA strand can be transcribed into double-stranded cDNA, which can then be inserted into prokaryotic vectors.

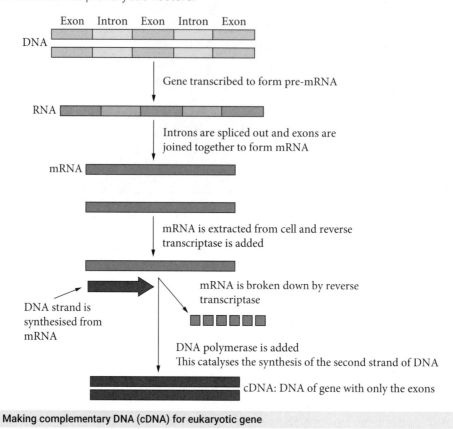

Making complementary DNA (cDNA) for eukaryotic gene

1.2.2 CRISPR-Cas9

CRISPR, pronounced 'crisper', is an acronym for Clustered Regularly Interspaced Short Palindromic Repeats. It is a short sequence of DNA that is used as a target for CRISPR-associated nucleases (Cas nucleases).

The CRISPR-Cas9 composite includes the endonuclease Cas9, which is able to cut DNA by catalysing the hydrolysis of the phosphodiester backbone of DNA, and a single guide RNA (sgRNA) sequence, which is used to guide Cas9 to the target DNA.

The function of CRISPR-Cas9 in bacteria

CRISPR-Cas9 is naturally occurring in bacteria and evolved as a defence against invading viruses. The Cas9 endonuclease is able to cut the invading viral DNA by recognising DNA sequences from the guide RNA. However, it is important that it only cuts the viral DNA and not the bacteria's genome. In order to protect its own DNA, it will only cleave at sites where a **protospacer adjacent motif (PAM)** is present. A protospacer is a 2–6 base sequence that immediately follows the target sequence for Cas9. In the bacteria *Streptococcus pyogenes*, Cas9 recognises the PAM NGG, where N can be any nucleotide followed by two guanine bases. Cas9 will cut where this sequence is found directly following the gene of interest, therefore protecting its own genome (self) that does not have the PAM sequence, while at the same time cutting viral DNA (non-self).

The application of CRISPR-Cas9 in editing an organism's genome

Scientists have manipulated this technique for gene editing. They can edit genes by synthesising a single guide RNA (sgRNA) that targets a specific section on the DNA. Cas9 will target this sequence if the correct PAM is present and cleave the DNA. Scientists have even manipulated the Cas9 protein to recognise other PAMs and not just NGG. It is thought the PAM destabilises the DNA and allows the sgRNA to be inserted. Cas9 will not function unless the correct PAM is present.

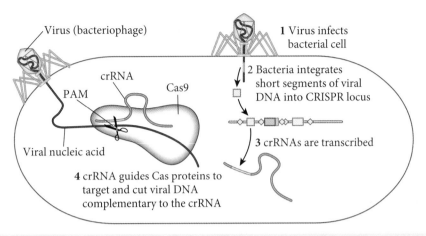

CRISPR-Cas9 in a bacterial defence system and how it can be applied to modify DNA

Once Cas9 has bound to the specific section of DNA and has cut it, the gene sequence can then be edited in a number of ways, including gene knock-in (also known as 'gene editing') or gene knock-out (also known as 'gene silencing').

Gene knock-in includes the gene of interest and the use of a **reporter gene**. A reporter gene is expressed under the control of a promoter gene. For example, when the promoter gene is switched on, the reporter gene will express a fluorescent green protein that glows when exposed to ultraviolet light. The expression of the green fluorescence can be measured in cells and tissue to show the location and whether the gene of interest is expressed. Conversely, a reporter gene can also inhibit the gene of interest, which makes the knock-in organism also a knock-out organism for that gene!

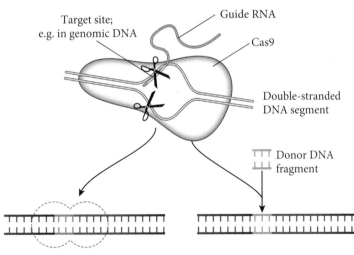

Gene knock-out, deleting a gene; and gene knock-in, inserting a gene

The CRISPR-Cas9 form of gene editing is now the most efficient and precise method of gene editing. It is low-cost and fast compared to other techniques.

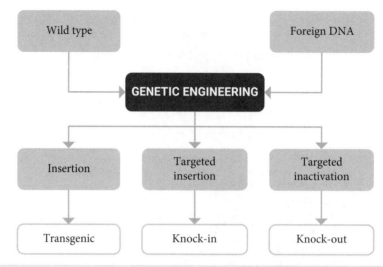

Using CRISPR for genetic engineering to target insertion or inactivation of specific genes

1.2.3 Amplification and sorting of DNA fragments

Polymerase chain reaction (PCR)

The **polymerase chain reaction (PCR)** is a three-step process involving a sequence of DNA that is amplified, or copied repeatedly, to produce large quantities of the DNA sequence. This is particularly useful for:

- application of small amounts of fossil DNA
- making copies of genes for insertion into other cells
- amplification for forensic and genetic testing
- making copies of artificially mutated DNA for further study.

PCR requires the DNA sample, primers, free nucleotides, Taq polymerase, and a buffer to keep the pH environment stable. All the ingredients are inserted into a PCR tube and placed in a PCR machine, where the following steps are repeated.

1 Denaturation (95°C): The DNA to be amplified is collected and separated into its two complementary strands by heating at high temperatures to break the hydrogen bonds between bases.

2 Annealing (50–60°C): The DNA is cooled and primers attach to the DNA and promote the replication process from the point of attachment.

3 Extending (72°C): The DNA polymerase known as **Taq polymerase** is an enzyme isolated from a bacterium, *Thermus aquaticus*. It used here because it can withstand the high temperatures without becoming denatured. Taq polymerase extends each strand beyond the primer, adding the free nucleotides to replicate each DNA strand. Each time this phase is completed, it will double the quantity of DNA.

PCR is extremely sensitive. Any contamination in the original sample will be copied over and over again, as will any mutation that occurs in the copying process. Extreme care is needed in a sample preparation and the process, particularly temperature, needs to be tightly controlled.

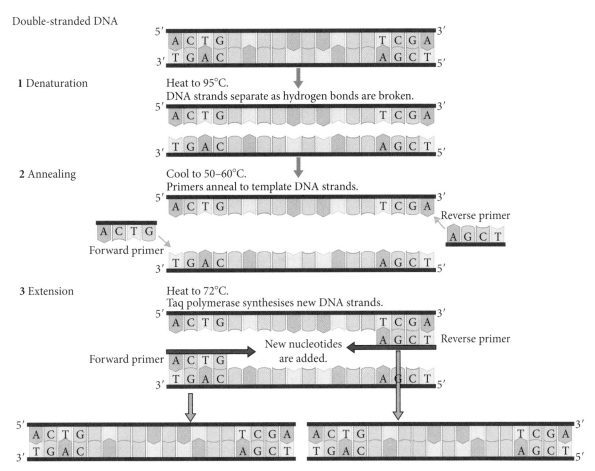

Double-stranded DNA

1 Denaturation

Heat to 95°C.
DNA strands separate as hydrogen bonds are broken.

2 Annealing

Cool to 50–60°C.
Primers anneal to template DNA strands.

Reverse primer

Forward primer

3 Extension

Heat to 72°C.
Taq polymerase synthesises new DNA strands.

Reverse primer

New nucleotides are added.

Forward primer

Cycle is repeated many times.

The steps of PCR

Gel electrophoresis

Gel electrophoresis is a process where nucleic acids are separated by molecular size. An electric current causes the DNA fragments to travel through the gel at different speeds. The ancient Greek word meaning 'to travel' is *phoresis*.

The gel is made of **agarose**, which is submerged in a **buffer** solution to maintain pH. The gel has **wells** across the top where the DNA is inserted. DNA is an acid, but because of the phosphate groups along the sugar–phosphate backbone, the DNA molecule is negatively charged. An electric charge is applied with the negative electrode near the DNA wells and the positive electrode at the opposite end. The DNA is attracted to the positive end, so it will move toward the positive electrode. This allows DNA fragments to be separated according to size; smaller fragments are able to fit through the pores of the gel easily and so can travel further through the gel. Large fragments cannot move as quickly.

Interpreting gels

A standard of known sizes, known as a **ladder**, must be used for each gel, as the length of DNA travelled will depend on the voltage, time in gel, buffer concentration and the gel composition.

In order to see the DNA and the ladder clearly in the gel, it must be treated first. It may be stained with a chemical that makes the DNA fluoresce under ultraviolet light, and photographed to show all the DNA bands. These bands can be compared to the ladder to enable comparison of the length of the fragments. The fragments are usually measured in **kilobases** (kb).

DNA profiles

Gel electrophoresis can be used for DNA profiling, which is where individuals can be identified by their DNA sequences. To do this, sections of non-coding regions of DNA, which form **satellite DNA**, can be removed to form fragments using restriction enzymes. Satellite DNA consists of variable lengths of repeating DNA called **short tandem repeats** (STRs, or microsatellites), usually 1–2 base pairs (bp) long, or variable region tandem repeats (VRTRs, or minisatellites), 10–100 bp long. Short tandem repeats occur when two or more

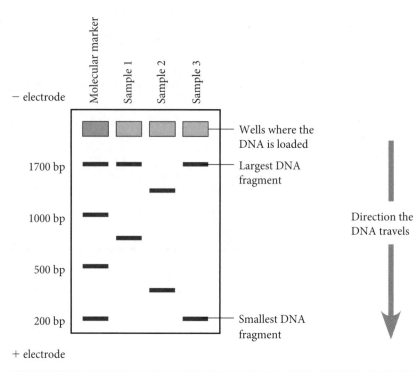

A gel electrophoresis set-up, showing the movement of DNA fragments through the gel toward the positive electrode

nucleotides are repeated and the repeats are next to each other. The individual will have a unique number of repeats but the same sequence, so they will produce unique fragment profiles based on the size of the sequence.

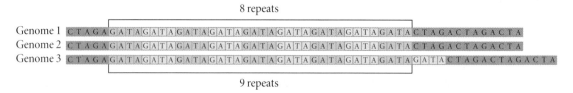

The number of tandem repeats is unique to the individual, so these can be used in DNA profiling

Applications of DNA profiling

DNA profiling can be used for paternity testing, forensic testing, genealogical and medical research. In order to compare DNA, a sample must first be taken from blood, semen, saliva, hair etc. Restriction enzymes are used to generate fragments that can then be amplified using PCR. The fragments are separated based on molecular size using gel electrophoresis. The movement of the fragments can then be interpreted.

When interpreting paternal parentage, you must remember that the child will inherit half of their DNA from their mother and half from their father, so they will express a combination of both parents' DNA. In the below diagram, you can see the STR alleles from a family that include the alleles from both parents, and each child who has inherited one allele from each of their parents.

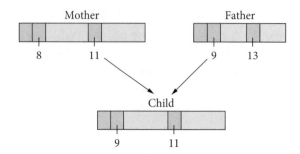

A short tandem repeat locus showing the inheritance of alleles from parents

In the case of a criminal investigation, the DNA from a sample must be a complete match for the DNA of the suspect in order to convict them of a crime.

Genotypes can also be tested, as variable regions may have different alleles. Two bands would indicate a heterozygous genotype and one band would indicate a homozygous genotype. The differences in the sizes of fragments is known as **restriction fragment length polymorphisms (RFLPs)**, which results in the unique band pattern.

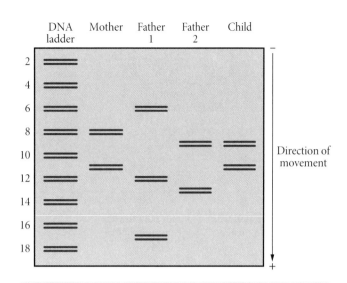

Gel electrophoresis results determining that the biological father of the child is Father 2

1.2.4 Use of recombinant plasmids as vectors to transform bacterial cells

Type 1 diabetes requires patients to self-inject with insulin in order to regulate their blood sugar levels so that cells can take up the glucose available for cellular respiration. Up until the 1980s, the only source of insulin was from the pancreas of pigs and cows. This had its drawbacks, including stimulating an allergic response in the patient as well as other ethical and health concerns. Human insulin is now used and is produced by

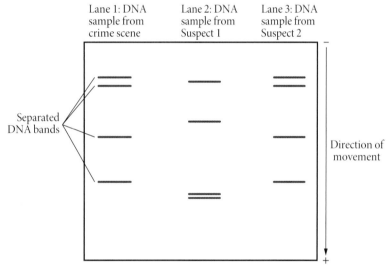

Gel electrophoresis comparing a DNA sample from a crime scene to the DNA of two suspects. Suspect 2 is a perfect match, indicating that this suspect was at the scene of the crime.

genetically engineering bacteria to produce insulin through the use of **recombinant plasmids**.

Plasmids are double-stranded circular DNA found in prokaryotes. They carry DNA that is additional to the circular or chromosomal DNA, and they replicate independently. Plasmids often carry genes that are not essential for life but do offer advantages for survival, including antibiotic resistance.

The ability for antibiotic resistant genes to be found in plasmids can be useful for scientists because antibiotic resistant genes can be placed into plasmids at the same time as a gene of interest, transforming the bacteria. If the transformation has worked, the bacteria will be resistant to an antibiotic, which can then be used in what is known as antibiotic selection, quickly allowing scientists to determine whether bacterial transformation has taken place.

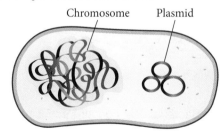

Plasmids are double-stranded DNA fragments separate to the chromosomal DNA.

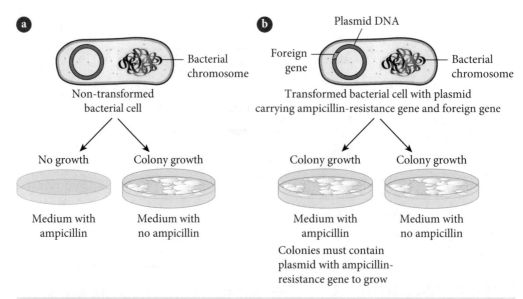

The process of antibiotic selection to screen for transformed bacteria. **a** Non-transformed bacteria cannot grow on media supplemented with ampicillin, but grow well on normal media. **b** Transformed bacteria can grow equally well on either medium. The plate without ampicillin provides a positive control condition.

Plasmids can be transferred to one another through **conjugation** and during replication via binary fission, where the bacteria will copy their plasmids so that the daughter cell receives all the same material that is contained in the parent cell.

Many copies of a gene may be made by inserting the gene into bacteria. For example, the insulin gene is inserted into a plasmid, forming a recombinant plasmid.

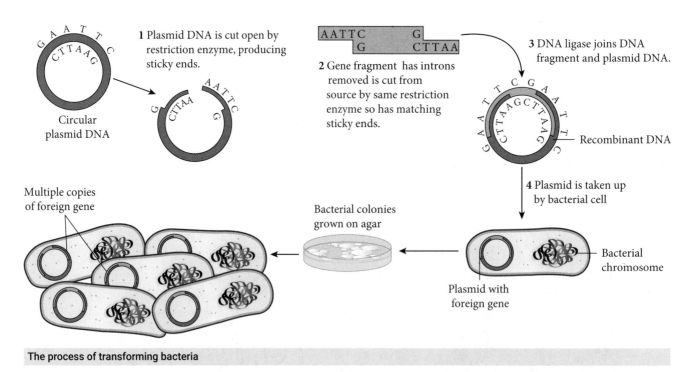

The process of transforming bacteria

Bacterial DNA contains exons but does not contain introns so all introns need to be removed prior to insertion of the DNA into the plasmid. Now the plasmid can be used as a **vector**, or genetic taxi, which transfers genetic material into a cell.

Human insulin has quaternary structure being made up of two separate polypeptide chains: A chain and B chain which are joined together by disulfide bonds to form a functional insulin molecule. Each chain is coded for by a different gene.

To make human insulin the two genes of interest are obtained using a technique that removes introns. Bacterial DNA does not contain introns, only exons, and does not have the machinery to splice them out. Leaving introns in the DNA will cause the genes not to be expressed. The two genes now travel separately and are grown in two separate cultures.

The production of both A chain and B chain are similar as described below.

The genes are prepared using restriction enzymes (EcoR1 and BamH1) that produces sticky ends. They are mixed with DNA ligase and *E. coli* plasmids which have been cut with the same restriction enzyme leaving the same sticky ends. These bacterial plasmids undergo further recombination to take up a modified gene for the protein β-galactosidase (*lacZ* gene) which is inserted next to the gene for the insulin protein. This is important to assist scientists in detecting successful gene insertion using a technique called blue-white screening. If the transformed bacteria are grown on agar plates, those colonies that grow white contain bacteria that have taken up the plasmid but without the gene of interest, those colonies that grow blue contain bacteria that have taken up the plasmid with the gene of interest. This can be used to separate the bacteria to grow only those with the gene of interest. Once produced, the insulin protein is removed from the protein β-galactosidase.

When two different proteins are produced (A chain and B chain), they are extracted from the bacteria, purified and mixed to form functional insulin.

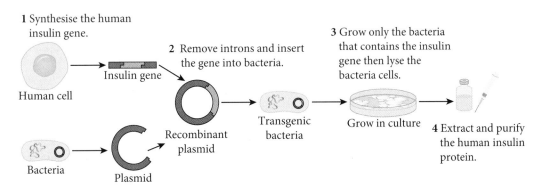

The steps involved in producing insulin gene for A chain in bacterial cells

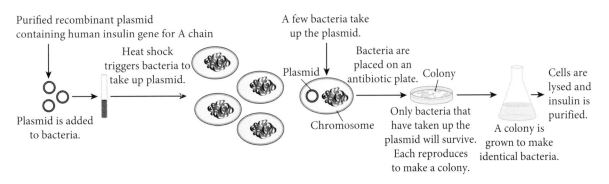

1.2.5 Use of genetically modified transgenic organisms in agriculture

Genetically modified organisms (GMO) have had their genes manipulated by either silencing a gene or turning on a gene to control expression. These genes are already present in the organism and the goal is to create a more favourable outcome.

Transgenic organisms have been genetically modified by having a particular gene inserted into their genome, which has been introduced from a different species. This allows the organism to express proteins that were not previously part of their genome.

No technology is without implications for the way we live, how we think, the environment or the future, and there are several ethical factors to consider when using genetically modified organisms in agriculture.

Genetic technology	Example	Ethical considerations
Transgenic: Insertion of genes into crop plants so they gain desirable features such as disease resistance, firmer fruit or resistance to herbicides	A bacterial gene inserted into crop plants that makes the plants resistant to the herbicide RoundUp®. Crops are sprayed with the herbicide, which kills the weeds but leaves the crop plants unaffected Mildew-resistant potatoes, so farmers don't have to treat their fields after heavy rain, saving time and money Bt corn, which produces its own pesticide	**Environmental:** It is possible that genes from GM crop plants will cross into non-GM varieties via dispersal of pollen/and or seeds. It is possible that the genes will also spread to other species via bacterial or viral infection. On a positive note, decreased use of herbicides and pesticides is positive for the environment. **Evolution:** Greater pressure for farmers to use higher-yielding or more resistant crop varieties from the gene pools of crop plants, thus reducing diversity. **Ethics:** Is it right that corporations hold ownership rights over crops that been grown for centuries, just because they developed one genetic variety? Are these varieties squeezing out traditional farming in poorer communities, exacerbating poverty in developing economies? If we have the technology, should we be helping to improve the lives of others? Is this seen to be 'playing god' and should we instead let nature take its natural course?
GMO: Modification through the silencing or switching on of genes to control expression to produce desirable traits	FLAVR SAVR tomatoes have been modified to remove genes that are responsible for the softening of fruit, so tomatoes stay fresh for longer Silencing the gene responsible for the production of a compound that causes apples to brown when they experience stress	**Evolution:** Many of the foods we eat today have been genetically modified through genetic technology or selective breeding. Every banana you eat is a clone and by reducing variation within the plant species, they are more at risk of disease and extinction.

Glossary

active site The region of an enzyme molecule that is the site for attachment of molecules of substrate

active transport The transport of molecules across the plasma membrane against a concentration gradient; requires the cell to expend energy in the form of ATP

agarose A carbohydrate, extracted from seaweed, which is heated and then cooled to form a gel used for separating DNA fragments according to size in gel electrophoresis

alpha helix (α-helix) A type of secondary protein structure in which the polypeptide chain folds into a tight coil

amino acid A small molecule that is the building block of protein; all amino acids contain an amino group (NH_2) and a carboxyl acid group (COOH)

amino acid sequence The linear order or arrangement of amino acids in the primary structure of a polypeptide

anabolic reaction A reaction in which the cellular reaction does not require oxygen; occurs in the cytosol of cells

anticodon The three nucleotides in tRNA that bind to the complementary codon in mRNA according to base-pairing rules, resulting in the addition of a specific amino acid to the polypeptide chain

ATP (adenosine triphosphate) A portable energy source that moves about the cell to sites in the cell where energy is required; ATP breaks down to form ADP (adenosine diphosphate) and inorganic phosphate (P_i), and releases energy together through condensation polymerisation forming strong covalent bonds

attentuation A mechanism for reducing the expression of the trp operon when levels of tryptophan are high by preventing completion of transcription

beta pleated (β-pleated) sheet A type of secondary protein structure in which segments of the polypeptide chain bond side by side into a flattened assembly

biochemical reaction A change of one molecule into a different molecule, which takes place inside a cell; often controlled by enzymes

blunt end the end of a DNA fragment that is created following cleavage by a restriction enzyme that cuts DNA at the same position on both strands

buffer A substance that maintains the pH of a solution

catabolic reaction A reaction in which there is a net gain or release of energy

cell wall A rigid structure surrounding the cell membrane that aids in protecting the cell and supporting it structurally; found in plants, fungi and

A+ DIGITAL FLASHCARDS
Revise this topic's key terms and concepts by scanning the QR code or typing the URL into your browser.

https://get.ga/a-biology-vce-u34

bacteria

cellular structure The components that make up a cell, including organelles

chaperone An enzyme that promotes the bending and folding of a polypeptide chain to produce a protein's tertiary structure

chloroplast An organelle found in plants and green algae, which is the site for photosynthesis

cloned The state in which cells, tissues or whole organisms with identical DNA have been duplicated

codon A group of three nucleotides in mRNA that specifies an amino acid

competent Describes the state when cells have been manipulated so they are more able to take up foreign DNA from their environment

concentration gradient When two areas have a difference in concentration of a substance; the larger the difference, the steeper the gradient

condensation polymerisation reaction The formation of polymers, such as peptides and carbohydrates, by a reaction that involves the release of water molecules; also known as condensation polymerisation

conjugation When two bacteria come into direct contact and one transfers genetic material to the other

covalent bond A chemical bond in which atoms share electrons

cytoplasm The cell membrane, organelles and cytosol together; all the cell except the nucleus

cytosol The semi-fluid matrix in which cellular organelles are suspended

degenerate (triplet code) A genetic code in which one amino acid may be coded for by more than one codon; also known as redundant triplet code

deoxyribonucleic acid (DNA) The information molecule that is the basis of an organism's genetic material

diffusion The tendency of particles of gases, liquids and solutes to disperse randomly and fill the available space

DNA ligase An enzyme that catalyses the bonding between two strands of DNA

DNA polymerase The enzyme that catalyses the bonding of nucleotides to form new strands of DNA

endocytosis The process of engulfing substances contained within a vesicle in order to bring them inside a cell

endonucleases A class of enzymes that catalyse the breaking of the bond between the phosphate and the sugar of the nucleic acid backbone of DNA

enzyme A specific protein catalyst that increases the rate of a biochemical reaction within the cell by lowering the amount of energy required for the reaction to proceed

eukaryotic Describes a complex type of cell with a nucleus and membrane-bound organelles

exocytosis The process where substances formed within the cell are transported in vesicles and fuse with the plasma membrane in order to release contents into the extracellular environment

exons Parts of a gene that contain the base sequences that code for the amino acids needed to build a peptide chain

facilitated diffusion The diffusion of molecules and ions from an area of high concentration to an area of low concentration; diffusion occurs through channels provided by embedded protein molecules

fibrous proteins Insoluble, strong proteins that usually provide strength and protection to cell structure; examples include collagen, and keratin found in hair and nails

flagella (singular: flagellum) A whip- or tail-like structure that aids in movement of a cell

gel electrophoresis A method used to separate nucleic acids by molecular size

gene A segment of DNA that determines the structure of a protein

gene expression The process by which the information in a gene is turned into a polypeptide

gene regulation The process by which gene expression is switched on or off

globular proteins Water-soluble, spherical proteins including many enzymes and haemoglobin

glycoprotein A molecule formed when a protein is chemically combined with a carbohydrate subunit

Golgi apparatus A collection of membranes that package and store substances into vesicles in preparation for their release from the cell

hydrogen bond A relatively weak bond formed by the attraction between polar molecules

hydrophilic Literally 'water loving'; describes molecules that have an affinity with water and are able to form hydrogen bonds with water molecules

hydrophobic Describes molecules that are insoluble in water and repel water molecules

hydrophobic interactions The relationship between water and low-water soluble molecules, where the low-water soluble molecules will avoid interacting with the water

hypertonic Describes a solution that has a higher solute concentration than the cell

hypotonic Describes a solution that has a lower solute concentration than the cell

introns Non-coding segments of DNA spliced between the exons; only found in eukaryotic DNA

ionic bond A chemical bond between ions (particles that have an electrical charge)

isotonic Describes a solution that has the same solute concentration as intracellular fluid

kilobase (kb) 1000 bases of DNA or RNA

ladder (in gel electrophoresis) A standard of known sizes of DNA fragments, used as a marker for molecular weight

lysosome A membrane-bound organelle containing digestive enzymes that can be used to destroy pathogens or break down unwanted material

messenger protein A signals that relays information between cells, tissues and organs to coordinate biological processes

messenger RNA (mRNA) A molecule copied from DNA that leaves the nucleus and travels to the ribosomes in the cytosol; it carries the instructions for protein manufacture

mitochondrion A membrane-bound organelle where aerobic respiration occurs to produce ATP for cellular functions

net gain of energy The difference between the energy expended and the energy gained from the reaction

nucleic acid A large, linear polymer built from nucleotide monomers bonded together; includes DNA and RNA

nuclear membrane A double membrane that surrounds the cell nucleus

nucleolus A small globular structure in the nucleus that synthesises ribosomal RNA

nucleus A membrane-bound organelle that contains the genetic material of eukaryotes

operator A segment of DNA adjacent to the promoter that a repressor protein can bind to; the combination of the operator and repressor prevents transcription

organelle Literally 'little organ'; a cellular structure that carries out a specific function; most are membrane bound

osmosis The net movement of water molecules across a semi-permeable membrane, from an area in which water is in high concentration (low solute concentration) to an area in which water is in low concentration (high solute concentration)

peptide bond A chemical bond that links two amino acids in a chain

phospholipid bilayer Two layers of phospholipids, with a hydrophilic (polar) phosphate head and a hydrophobic (non-polar) tail; the tails arrange themselves so they point inwards; the phosphate heads arrange themselves so they are on the exterior

plasma membrane A phospholipid bilayer that separates the inside of the cell from the external environment

plasmid A small, circular DNA structure independent of the chromosome in prokaryotic cells

plasmolysis The process in which water is lost from a cell, which causes the cell to shrink or contract

polymerase chain reaction (PCR) A process used to amplify or make many copies of a piece of DNA

primary structure (of a protein) A protein's amino acid sequence

prokaryotic Describes a simple type of cell that lacks a nucleus and membrane-bound organelles; the smallest living cell; includes bacteria, archaebacteria and eubacteria

promoter An upstream region of DNA recognised by RNA polymerase, which binds to the DNA in order to initiate transcription

protein channel A channel within a cell membrane that proves a pathway for water or polar molecules that cannot diffuse through the phospholipid bilayer

protein molecule A large class of biological molecule that is a polymer of amino acids; acts as an enzyme, antibody, receptor or transport molecule

proteome The complete complement of an organism's proteins

proteomics The study of an organism's proteins

protospacer adjacent motif (PAM) A short sequence of DNA following the DNA sequence targeted by Cas9 nuclease on the non-target strand

recombinant plasmid A small, circular, double-stranded molecule of DNA that has had DNA fragments or whole genes inserted into it

reporter gene A gene that is introduced specifically for the detection or measurement of gene expression

repression A mechanism for reducing the expression of the *trp* operon when levels of tryptophan are high by blocking transcription

restriction enzyme A protein that is able to cut DNA at a specific location; produced by bacteria

restriction fragment length polymorphisms (RFLPs) Differences within individuals in the length of DNA fragments cut by enzymes; the variations can be analysed and used to observe mutations in genes

reverse transcriptase An enzyme used to synthesise DNA (cDNA) from an RNA template

ribonucleic acid (RNA) A type of nucleic acid consisting of a single strand of nucleotides; has essential roles in protein synthesis

ribosomal RNA (rRNA) A type of non-coding RNA that associates with a set of proteins to form ribosomes

ribosomes Organelles found within all living cells that synthesise proteins

rough endoplasmic reticulum Endoplasmic reticulum with ribosomes attached

satellite DNA Short, repetitive nucleotide sequences that are found in non-coding regions of DNA

secondary structure (of a protein) The localised folding of a polypeptide chain when neighbouring amino acids bond to each other to form α-helices, β-pleated sheets or random loops

short tandem repeats Two or more nucleotides that repeat, and the repeated sequences are next to each other, forming a sequence; found in non-coding regions of DNA

simple diffusion The movement of molecules through the phospholipid bilayer down a concentration gradient

smooth endoplasmic reticulum Endoplasmic reticulum with no ribosomes attached

solute A substance that is dissolved by a solvent to create a solution

sticky end The end of a DNA fragment that is created following cleavage by a restriction enzyme that cuts DNA at different positions on each strand

sulfur A chemical element found in the amino acid cysteine, which forms strong disulfide bonds between two cysteine molecules

taq polymerase An enzyme from *Thermus aquaticus*, which is heat stable, and is used in DNA polymerase chain reaction to synthesise and amplify new strands of DNA

transfer RNA (tRNA) An adaptor molecule composed of RNA, which delivers specific amino acids during protein synthesis

transformed bacteria Bacteria that have taken up and express foreign genetic material from the environment

triplet A set of three nucleotide codes

unambiguous Describes the state in which each codon codes for one amino acid only, so there is no misunderstanding

vacuole A membrane-bound organelle filled with fluid; used for storage or digestion

vector A carrier of disease, or a carrier of a piece of foreign DNA into a cell

well (in gel electrophoresis) An indent on the agarose gel where the DNA is loaded; wells should always be at the negatively charged end

Revision summary

In the table below, provide a brief definition of each of the key concepts. This will ensure that you have revised all the key knowledge in this Area of Study in preparation for the exam.

What is the role of nucleic acids and proteins in maintaining life?	
Nucleic acids as information molecules	
Structure of DNA	
Three main forms of RNA (mRNA, rRNA and tRNA)	
Genetic code as a universal triplet code that is degenerate	
Steps in gene expression, including transcription, RNA processing in eukaryotic cells and translation by ribosomes	
The structure of eukaryotic and prokaryotic genes, exons, introns and promoter and operator regions	
Gene regulation and the prokaryotic *trp* operon as an example	
Amino acids as the monomers of a polypeptide chain	
The hierarchical levels of a functional protein	
Proteins make an organism's proteome	

What is the role of nucleic acids and proteins in maintaining life?	
Enzymes are proteins that are biological catalysts	
The protein secretory pathway	
DNA manipulation techniques and applications	
Polymerase enzymes	
DNA ligase	
Endonucleases	
CRISPR-Cas9 in bacteria	
CRISPR-Cas9 application in gene editing	
Amplification of DNA using PCR	
Gel electrophoresis in sorting DNA fragments	
DNA profiling using gel electrophoresis	
Bacterial transformation using recombinant plasmids	
Genetically modified and transgenic organisms in agriculture	

Exam practice

The relationship between nucleic acids and proteins

Solutions start on page 212.

Multiple-choice questions

Question 1 ◐▨▨

Nuclear DNA (deoxyribonucleic acid) is a double-stranded nucleic acid molecule. In a double-stranded DNA molecule the

A number of guanine bases is equal to the number of cytosine bases.

B number of thymine bases is twice the number of cytosine bases.

C numbers of all four bases (adenine, thymine, guanine and cytosine) are equal.

D numbers of adenine and uracil bases are equal.

Question 2 ◐▨▨

A single DNA nucleotide contains

A ribose sugar, phosphate and thymine.

B deoxyribose sugar, phosphate and uracil.

C ribose sugar, phosphate and uracil.

D deoxyribose sugar, phosphate and thymine.

Question 3 ©VCAA | VCAA 2015 SA Q4 | ◐◐▨

The diagram below represents part of a DNA molecule.

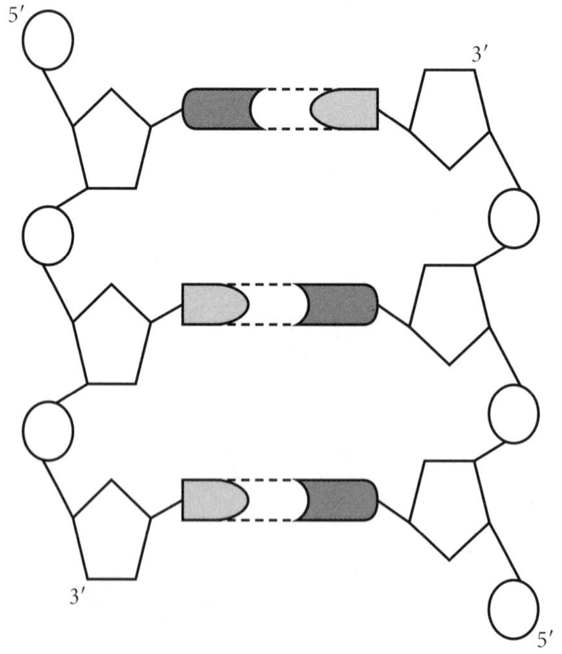

A feature of DNA that can be seen in the diagram is

A the anti-parallel arrangement of the two strands of nucleotides.

B the process of semi-conservative replication.

C its ribose sugar–phosphate backbone.

D its double-helix structure.

Question 4 ⬤⬤⬤

The codon GGA codes for the amino acid glycine. It is reasonable to conclude that

A the DNA segment that acts as the template for this codon is CCU.

B the tRNA anticodon with which this codon pairs will have the sequence CCT.

C the codon GGA may also code for other amino acids.

D codons other than GGA may code for glycine.

Question 5 ⬤⬤○

During RNA processing in the nucleus

A introns are retained and exons are removed.

B a poly-A tail is added to the 5′ end of the pre-mRNA.

C exons are retained and introns are removed.

D the positions of introns and exons are rearranged.

Question 6 ⬤⬤○

The following molecule of tRNA bears the amino acid arginine.

Which DNA triplet, in a template strand, corresponds to the tRNA molecule in the diagram?

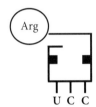

A AGG

B TGG

C TCC

D ACC

Question 7 ○⬤○

The structural order within a gene on prokaryotic DNA is

A promotor, operator, genes.

B genes, operator, promotor.

C promotor, genes, operator.

D operator, promotor, genes.

Question 8 ⬤⬤○

The promotor region of a gene is needed to

A promote the translation of the gene into a functioning protein.

B allow RNA polymerase to recognise the region and bind so the process of transcription can occur.

C identify the introns and exons across a piece of DNA for splicing.

D enable a repressor protein to bind so that the gene is not transcribed.

Question 9 ⬤⬤○

A prokaryotic cell found itself in an environment lacking tryptophan (an amino acid). This resulted in

A the *trp* operon being turned off by a repressor protein because the production of these enzymes was not required.

B an increase in cell mass because it could not excrete the by-products of its metabolism.

C the *trp* operon being free of repressors so that the enzymes required to synthesise tryptophan could be produced.

D no significant changes to the metabolic processes occurring in the cell.

Question 10 ●●●

This diagram shows

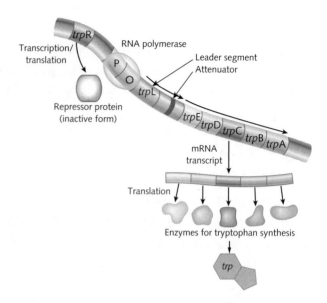

A the production of repressor proteins to reduce the synthesis of the enzymes to make tryptophan on the *trp* operon.

B in low levels of tryptophan, *trp* repressor proteins can't bind to the DNA and the enzymes required for the production of tryptophan can be transcribed and translated.

C DNA polymerase can bind to the *trp* operon so that replication of this operon can occur.

D the *trp* operon can be translated by RNA polymerase and then transcribed into tryptophan.

Question 11 ●●●

In the *E. coli* bacteria, attenuation is one method of gene regulation that prevents the production of energy-expensive tryptophan being made in the cell. Attenuation is made possible because

A the repressor protein is blocking the transcription of the five structural genes.

B transcription and translation occur simultaneously in the same part of the cell.

C sections 1 and 2 of the leader region form a hairpin loop.

D adenine and uracil bases are bonded by triple hydrogen bonds.

Question 12 ●●●

Which of the following is true of a protein's primary structure?

A It is formed by hydrogen bonds between neighbouring amino acids.

B It may include a combination of alpha helices and beta-pleated sheets.

C It involves peptide bonds between neighbouring amino acids.

D It is determined by the properties of the amino acid R groups.

Question 13 ●●●

The diagrams below show various monomers of biomolecules being joined together.

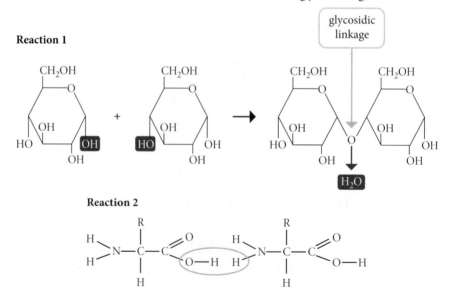

These reactions are

A hydrolysis reactions, because water is involved.

B energy-releasing reactions.

C catalysed by DNA ligase.

D condensation reactions, because a water molecule is released.

Question 14

The tertiary structure of a protein involves

A covalent bonds only between amino acids.

B hydrogen bonds between the hydrogen and oxygen atoms in the backbone of the molecule, forming alpha helices and beta-pleated sheets.

C ionic, covalent and hydrogen bonds plus hydrophobic interactions between the R side chains of the amino acids.

D multiple protein chains bonded together to form a functioning protein.

Question 15

All of the molecules below are part of the human proteome except

A the enzyme that catalyses the hydrolysis of carbohydrates.

B the structural protein collagen, found in connective tissue.

C haemoglobin, which is responsible for carrying oxygen around the body.

D glycogen, which stores glucose for energy.

Question 16 ©VCAA VCAA 2013 SA Q8

Consider the following reaction in which substrate molecule R and substrate molecule S are converted into product molecule T and product molecule U.

R and S → T and U

The graph on the right shows the energy available in the molecules against time.

Based on the information in the graph, a correct conclusion would be that

A this is an anabolic reaction.

B the reaction would release energy.

C the value of the activation energy for the reaction is shown by X.

D product molecules T and U have less energy than substrate molecules R and S.

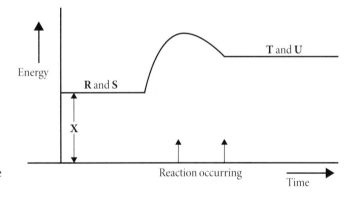

Use the following information to answer Questions 17 and 18.

Alanine aminotransferase (ALT) is an enzyme produced by the liver. It converts alanine, an amino acid found in proteins, into pyruvate, an important molecule in glycolysis (cellular energy production). Most ALT that the liver produces stays within the organ. However, when the liver is damaged or inflamed, it may release ALT into the bloodstream. In healthy individuals, ALT levels in the blood are low.

Question 17

In pancreatic cells, the site of synthesis of ALT is the

A nucleus.

B ribosomes.

C Golgi body.

D smooth endoplasmic reticulum.

Question 18 ⬤⬤◯

If high levels of ALT are seen in the bloodstream, it exits the liver cells by

A exocytosis.

B endocytosis.

C phagocytosis.

D breakage in the plasma membrane.

Question 19 ⬤⬤⬤

Pancreatic cells will incorporate radioactively labelled amino acids into proteins. Proteins can therefore be labelled and tracked within a cell. If we were tracking a pancreatic enzyme that was to be secreted, the most likely pathway of movement of this protein in the cell would be

A endoplasmic reticulum to Golgi apparatus to nucleus.

B Golgi apparatus to endoplasmic reticulum to lysosome.

C nucleus to endoplasmic reticulum to Golgi apparatus.

D endoplasmic reticulum to Golgi apparatus to secretory vesicles.

Short-answer questions

Question 20 (11 marks) ⬤◯◯

Below is a short sequence of the template (anti-sense) strand of a double-stranded segment of DNA:

CGA TGA CAT

a Copy and complete the table below with the sequences that would be found on each of the different types of RNA (excluding rRNA) and their respective roles in the synthesis of a protein product. 6 marks

Nucleic acid	Sequence of nucleotides	Function of the molecule
DNA	CGA TGA CAT	
mRNA		
rRNA	-----------------------------	
tRNA		

b On which type of nucleic acid above would you find

i a codon? 1 mark

ii an anticodon? 1 mark

c Using the table below, determine the sequence of amino acids translated from this DNA sequence. 1 mark

Second base

		U	C	A	G	
First base	**U**	UUU ⎤ Phe UUC ⎦ UUA ⎤ Leu UUG ⎦	UCU ⎤ UCC ⎥ Ser UCA ⎥ UCG ⎦	UAU ⎤ Tyr UAC ⎦ UAA Stop UAG Stop	UGU ⎤ Cys UGC ⎦ UGA Stop UGG Trp	U C A G	**Third base**
	C	CUU ⎤ CUC ⎥ Leu CUA ⎥ CUG ⎦	CCU ⎤ CCC ⎥ Pro CCA ⎥ CCG ⎦	CAU ⎤ His CAC ⎦ CAA ⎤ Gln CAG ⎦	CGU ⎤ CGC ⎥ Arg CGA ⎥ CGG ⎦	U C A G	
	A	AUU ⎤ AUC ⎥ Ile AUA ⎦ AUG Met/ Start	ACU ⎤ ACC ⎥ Thr ACA ⎥ ACG ⎦	AAU ⎤ Asn AAC ⎦ AAA ⎤ Lys AAG ⎦	AGU ⎤ Ser AGC ⎦ AGA ⎤ Arg AGG ⎦	U C A G	
	G	GUU ⎤ GUC ⎥ Val GUA ⎥ GUG ⎦	GCU ⎤ GCC ⎥ Ala GCA ⎥ GCG ⎦	GAU ⎤ Asp GAC ⎦ GAA ⎤ Glu GAG ⎦	GGU ⎤ GGC ⎥ Gly GGA ⎥ GGG ⎦	U C A G	

d The genetic code is described as degenerate and universal. What do these two terms mean? Provide an example in your explanation.   2 marks

Question 21 (4 marks) ©VCAA VCAA 2019 SB Q1 ●●●

Diagrams of two molecules that are required for the production of proteins within a cell are shown below.

Molecule 1

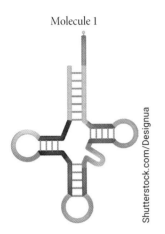

Shutterstock.com/Designua

Molecule 2

Shutterstock.com/PannaKotta

a Copy and complete the table to describe two differences between the monomers of the two molecules.   2 marks

	Molecule 1	Molecule 2
Difference 1		
Difference 2		

b Ten amino acids that form part of a protein are shown below.

– Phe – Val – Asn – Gln – His – Leu – Cys – Gly – Ser – His –

The section of an RNA molecule found in the nucleus of the cell associated with the translation of these 10 amino acids was found to contain more than 300 monomers. Explain how there can be more than 300 monomers in this section of the RNA molecule but only 10 amino acids translated.   2 marks

Question 22 (6 marks) ●●○

Most organisms have a set of genes, called heat shock genes, that encode proteins known as heat shock proteins. Heat shock proteins usually assist in the folding of protein molecules and repairing damaged proteins. Heat shock genes are activated under conditions of heat stress and are controlled by a transcription factor called heat shock transcription factor (transcription factors are proteins that control the rate of transcription).

In order to be expressed, heat shock genes must be transcribed and translated. Outline the process and location of

a transcription.   3 marks

b translation.   3 marks

Question 23 (7 marks) ©VCAA VCAA 2015 SB Q7 ●●●

Glucocorticoid (GC) is a hormone in rats that binds to a receptor, as shown in the diagram below. The glucocorticoid–receptor complex (GCR-complex) moves into the nucleus and attaches to the DNA, causing transcription to begin.

GC signal transduction in rat pituitary gland cells

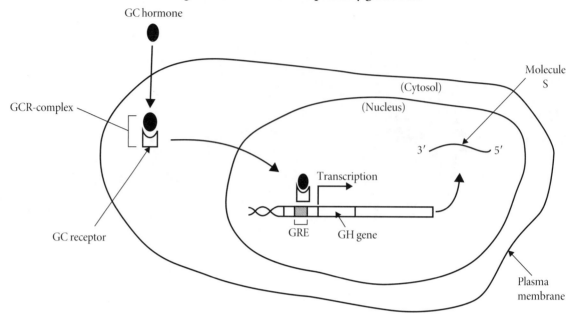

The location where the GCR-complex attaches to the DNA is called the glucocorticoid response element (GRE). The GRE is located approximately 250 base pairs upstream of the growth hormone (GH) gene. Following the attachment of the GCR-complex to the GRE, an enzyme catalyses the transcription of the gene.

a Name the enzyme that catalyses transcription.   1 mark

b Name the transcription product, molecule S, and describe the processing that molecule S undergoes before it exits the nucleus.   3 marks

c In the rat pituitary gland, GC stimulates the production of the growth hormone protein. However, in the rat liver, GC stimulates the production of the enzyme tryptophan oxygenase. Given that the genetic sequence is identical in all somatic rat cells, explain how the production of distinct proteins in different cell types could occur.   2 marks

d If a human gene is inserted into the DNA of rat pituitary gland cells, these genetically engineered cells can be used to produce human growth hormone. What characteristic of the genetic code enables a human protein, such as human growth hormone, to be made by rat cells?   1 mark

Question 24 (2 marks) ●●○

Explain how a repressor protein functions in the process of gene regulation.   2 marks

Question 25 (2 marks) ●○○

List **one** similarity and **one** difference between the structure of genes in eukaryotes and prokaryotes.   2 marks

Question 26 (1 mark) ●●○

Explain how eukaryotic organisms can produce multiple proteins from the one gene.   1 mark

Question 27 (2 marks) ©VCAA VCAA 2015 SB Q8a ●●●

Consider the template strand of a hypothetical gene, shown below. The exons are in bold type.

3' **TAC AAA** CCG GCC **TTT GCC AAA** CCC AAC CTA **AAT ATG AAA ATT** 5'

Note: The DNA triplet TAC indicates START and codes for the amino acid methionine that remains in the polypeptide. The DNA triplets ATC, ATT and ACT code for a STOP instruction.

a How many amino acids would be present in the polypeptide expressed by this gene?   1 mark

b An allele for this gene codes for a polypeptide with only five amino acids. This is caused by a mutation in one of the exons. This mutation is a result of one nucleotide change. By referring to the original sequence above, identify the nucleotide change that must have occurred to bring about this shorter polypeptide.   1 mark

Question 28 (8 marks) ©VCAA VCAA 2013 SB Q6 ●●●

The following diagram outlines various events that occur in cells when DNA is activated.

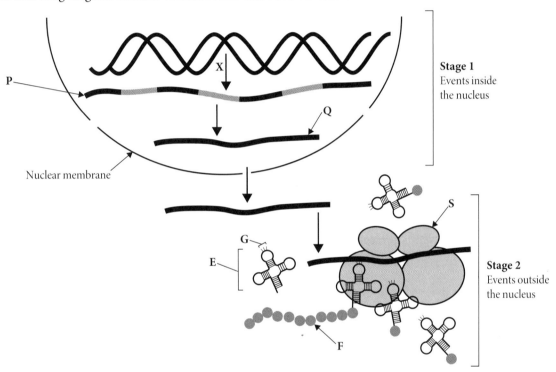

a **i** Outline events that occur during action X.   2 marks

   ii Note that structure P consists of two different kinds of components. What are these two components called and what happens to each component?   2 marks

b Describe the events occurring in stage 2, including the role of each of the structures S, F, E and G.   4 marks

Question 29 (6 marks) ●●●

Bacteria such as *E. coli* need the amino acid tryptophan to survive. *E. coli* can ingest tryptophan from its environment and can also synthesise it using enzymes that are encoded by five genes. These five genes are next to each other in what is called the tryptophan (*trp*) operon. If tryptophan is present in the environment, then *E. coli* does not need to synthesise it. A repressor protein coded for by *trp*R binds to the operator only after it has been activated by tryptophan. The following diagram is a simplified version of the *trp* operon when tryptophan is plentiful in the cell.

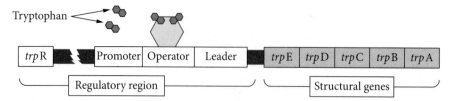

a   What is an operon?   1 mark

b   What type of gene is *trp*R?   1 mark

c   Which genes shown in the diagram code for the production of the enzymes necessary
    to synthesise tryptophan?   1 mark

d   In the situation shown in the diagram above, explain what would be the expected rate
    of transcription of the five structural genes.   1 mark

e   Explain how this operon would function in the absence of tryptophan.   2 marks

Question 30 (5 marks) ⬤⬤⬤

A study was done looking at the effect of mutations associated with the *trp* operon in *E. coli*. In Mutant X, the operator region of the operon has a mutation that changes the sequence of the nucleotides and thus alters its complementary relationship with other molecules so that, once bound, they cannot dissociate. Mutant Y has a point mutation that changes the shape of the tryptophan binding site on the repressor protein.

a   Will these mutations have any effect on the expression of the structural genes
    associated with the *trp* operon?   1 mark

b   Describe the impact of these mutations on the production of tryptophan when there
    is tryptophan present in the environment for both cells.   4 marks

Question 31 (6 marks) ⬤⬤◐

a   The diagram on the right represents the amino acid alanine.

    i   What **two** features of this molecule tell you that it is an
        amino acid?   1 mark

    ii There are 20 amino acids commonly found in human cells. In what way do
        the molecules of other amino acids differ from this molecule of alanine?   2 marks

b   The following diagram shows details of the secondary structure of part of a protein molecule.

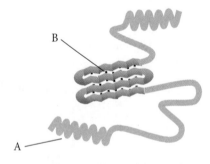

    i   Name the structures labelled A and B.   2 marks

    ii What type of chemical bonds are responsible for the formation of structures A and B?   1 mark

Question 32 (5 marks) ⬤⬤◐

a   What is an enzyme and how do enzymes catalyse reactions?   2 marks

b   With the aid of a diagram, describe a catabolic reaction with regards to the reactants
    and then products produced, and the changes in energy.   3 marks

Question 33 (4 marks) ⚫◦◦

Using the table below, describe the function of the different types of proteins and provide **one** example.   4 marks

Protein	Function	Example
Contractile proteins		
Hormonal proteins		
Structural proteins		
Transport proteins		

Question 34 (6 marks) ⚫⚫⚫

Lipase is an enzyme that catalyses the hydrolysis of fats (lipids). Lipases perform essential roles in digestion, transport and processing of dietary lipids (e.g. triglycerides, fats, oils) and are produced by pancreatic cells. Nucleic acids encode instructions for the synthesis of lipase in pancreatic cells.

a   Outline the steps involved in the translation of lipase during synthesis in pancreatic cells.   3 marks

b   After being synthesised, lipase is released from pancreatic cells via exocytosis. Name **three** different organelles directly associated with the transport of the synthesised lipase within or from pancreatic cells and state the role of each organelle in this process.   3 marks

DNA manipulation techniques and applications

Solutions start on page 218.

Multiple-choice questions

Question 1 ©VCAA VCAA 2019 SA Q39 ⚫⚫⚫

The diagram at right is a map of a bacterial plasmid showing ORI, the origin of DNA replication, and selected restriction endonuclease sites.

One plasmid was mixed with the restriction enzymes *Eco*RI, *Bam*HI and *Hinc*II. Which of the following shows the number of restriction sites that have been cut and the resulting number of DNA fragments produced?

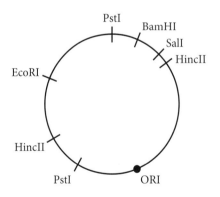

	Number of restriction sites cut	Number of DNA fragments produced
A	3	3
B	3	4
C	4	4
D	4	5

Question 2 ©VCAA VCAA 2016 SA Q36 ⚫⚫◦

During the preparation of the recombinant plasmids, researchers used the enzyme DNA ligase. The function of DNA ligase for this purpose is to

A   manufacture an antigen that will be recognised by T helper cells.

B   join the dystrophin gene to the plasmid DNA at complementary sticky ends.

C   clone the plasmid in order to produce enough plasmids to ensure effective treatment.

D   cut the DNA of the plasmid and the dystrophin gene in the same manner in order to produce matching sticky ends.

Use the following information to answer Questions 3 and 4.

Genetic engineers use restriction enzymes to cut DNA into smaller lengths. The recognition sequences of several restriction enzymes are shown in the table below. The symbol * denotes the restriction site (position of the cut).

Restriction enzyme	Recognition sequence (read in 5′ to 3′ direction)					
EcoRI	G *	A	A	T	T	C
	C	T	T	A	A	* G
HindIII	A *	A	G	C	T	T
	T	T	C	G	A	* A
AluI		A	G *	C	T	
		T	C *	G	A	
HaeIII		G	G *	C	C	
		C	C *	G	G	

Consider a length of double-stranded DNA with the sequence

5′ C T A G C T G A A T T C A A G G C C T C 3′
3′ G A T C G A C T T A A G T T C C G G A G 5′

Question 3

The restriction enzyme that would result in two double stranded DNA fragments with sticky ends would be

A EcoRI.

B HindIII.

C AluI.

D HaeIII.

Question 4

Which of the following statements is correct?

A If this segment was cut with HaeIII, the result would be four single-stranded fragments of DNA.

B If both EcoRI and HindIII were added to this fragment, the result would be two double-stranded DNA fragments.

C If this segment was cut with AluI, the result would be three double-stranded segments of DNA.

D If this segment was exposed to all four restriction enzymes, the result would be five double-stranded DNA fragments because each would cut the section once.

Question 5

The function of CRISPR-Cas9 in bacteria is

A to produce proteins to encase an invasion of viral DNA so it can be exocytosed.

B as a defence mechanism that cleaves viral DNA that enters the cell.

C to multiply plasmids that produce antiviral proteins.

D as an enzyme that speeds up the rate of bacterial replication (binary fission).

Question 6

The CRISPR-Cas9 technique has been used by researchers

A to replicate DNA during PCR.

B as a precursor to gel electrophoresis.

C as a way to clone genes in order to treat human diseases.

D as a way to edit genes so that they can be silenced, or genes can be inserted into specific places in a genome.

Use the following information to answer Questions 7 and 8.

Four samples of DNA were loaded into four different wells in lanes W, X, Y and Z. A standard ladder was loaded into the well in lane S. The results of gel electrophoresis are shown at the right.

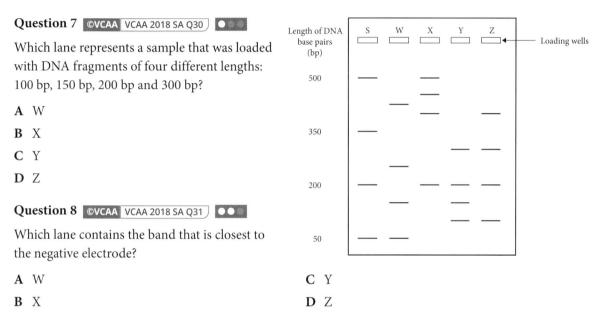

Question 7 ©VCAA VCAA 2018 SA Q30

Which lane represents a sample that was loaded with DNA fragments of four different lengths: 100 bp, 150 bp, 200 bp and 300 bp?

A W

B X

C Y

D Z

Question 8 ©VCAA VCAA 2018 SA Q31

Which lane contains the band that is closest to the negative electrode?

A W

B X

C Y

D Z

Question 9

The polymerase chain reaction (PCR) is a method used to

A amplify the number of copies of specific sections of DNA so that they can be detected using gel electrophoresis.

B identify a variety of pathogens with one set of primers.

C insert plasmids into bacteria so that they can be amplified within the rapidly dividing bacteria in order to increase the number of copies.

D cut DNA into smaller fragments so they can be inserted into plasmids.

Question 10

The diagram below represents a DNA molecule and the position of recognition sites along the molecule for three different restriction enzymes.

This DNA strand was digested using the restriction enzyme *Sal*I and then separated using gel electrophoresis. Using your knowledge of gel electrophoresis because there is no DNA ladder for comparison, which lane shows the most likely result of this process?

A Lane 1

B Lane 2

C Lane 3

D Lane 4

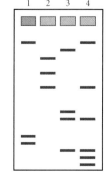

Question 11 ⬤⬤▢

Barmah Forest virus is a uniquely Australian pathogen that infects around 400 people each year. Humans that contract this disease have been bitten by mosquitos that have come into contact with the virus from infected marsupials or other infected humans. It can be detected using the polymerase chain reaction (PCR). Blood samples from patients are tested via PCR. The results will show

A a positive or negative result depending on whether the virus is present in the sample.

B a positive or negative result depending on whether the patient has developed antibodies to the virus.

C bands on a gel and the presence of a specific band indicates the presence of the virus.

D a positive result when the sample fluoresces under UV light because a plasmid has been used to identify the virus and produce a fluorescent protein.

Use the following information to answer Questions 12–14.

The production of dyes for food and cosmetics can be a costly and difficult process. So much so that manufacturers often rely on the artificial production of these dyes, which can have detrimental effects on the environment. However, in 2017, researchers were able to break the complex process of producing anthocyanins (purple and blue dyes) down into four steps using four different transformed bacteria. The process is outlined below.

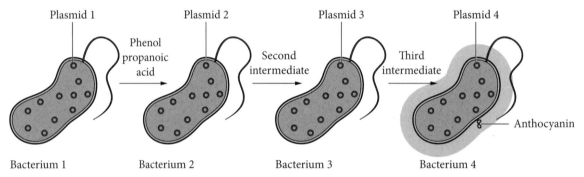

1   Bacterium 1 is transformed with a plasmid that contains a gene to produce an enzyme that uses glucose to produce easily absorbed intermediate products: phenylpropanoic acids.

2   Bacterium 2 is transformed with a plasmid containing a gene for enzymes to produce a second intermediate. This bacterium ingests the phenylpropanoic acids and uses them as reactants.

3   Bacterium 3 is transformed with a third plasmid containing genes for enzymes to use the second intermediate molecules from bacterium 2 to produce a third molecule.

4   Bacterium 4 is transformed with a fourth plasmid containing genes for enzymes to produce anthocyanin from the products from bacterium 3.

Question 12 ⬤▢▢

Anthocyanin is produced by bacterium

A 1.   **C** 3.

B 2.   **D** 4.

Question 13 ⬤⬤⬤

Why were four different transformed bacteria used in this process and not just one?

A The bacteria contain different restriction enzymes that they will use to produce the products required.

B There were too many steps involved and the bacteria could not be transformed with such a large number of genes if the products were still to be controlled by researchers.

C Due to the commercial nature of this process, four bacteria are needed to produce the large quantities of product required.

D Bacterium 1 will only accept plasmid 1, bacterium 2 will only accept plasmid 2 and so on.

Question 14 ⬤⬤⬤

Before this process can occur, the four different bacteria need to be transformed with the recombinant plasmids so that they can be used in this pathway.

In order to identify which bacteria are transformed with the appropriate plasmid, researchers could

A grow the bacteria on agar plates and look for the pigment anthocyanin, as all of the transformed bacteria will produce this product.

B grow the bacteria in a liquid broth, as all bacteria undergo the transformation process.

C make sure the plasmid contains another gene for antibiotic resistance and grow the bacteria on agar plates containing this antibiotic.

D place all four types of bacteria on one agar plate and only those producing anthocyanin will be transformed.

Question 15 ⬤◯◯

A genetically modified organism (GMO)

A has experienced a change in a single nucleotide in its DNA, resulting in a change in the conformational shape of the resulting protein.

B is an organism that has had its genome manipulated by either silencing a gene or turning on a gene to control the expression of a gene product.

C is where humans have decided on a desirable trait and have selectively bred the organism to improve the expression of those desirable traits.

D cannot be used to modify the genomes of other organisms.

Question 16 ⬤◯◯

A transgenic organism is an organism that has

A adapted to grow in a new environment.

B had its genome modified to include a gene from another individual of the same species.

C been genetically modified by having a particular gene inserted into its genome, which has been introduced from a different species.

D been exposed to a mutagen that causes mutations in its genome, resulting in different proteins being produced.

Question 17 ©VCAA VCAA 2019 SA Q37 ⬤⬤◯

Bt corn expresses a protein that acts as an insecticide.

Based on your knowledge and the data in the graph, what is a benefit of using Bt corn?

A More insecticide is used with Bt corn crops.

B Bt corn is cheaper to produce than non-Bt corn.

C Negative impacts on ecosystems could be reduced.

D Fewer farmers are predicted to plant Bt corn in the future.

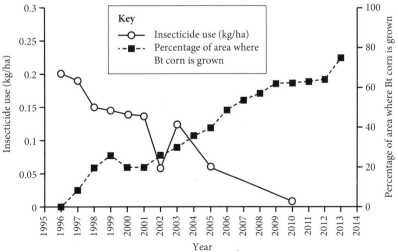

Bt corn uptake and insecticide use in cornfields

'Smarter pest control', special section, Science, vol.341, 16 August 2013, p. 731.
Reprinted with permission from AAAS.

Short-answer questions

Question 18 (5 marks)

Name of enzyme	Source	Recognition site and cleavage site	Nature of cut ends
*Eco*RI	*E. coli* RY13	5′- G|AATTC - 3′ 3′- CTTAA|G - 5′	Sticky
*Hind*III	*Haemophilus influenzae* Rd	5′- A|AGCTT - 3′ 3′- TTCGA|A - 5′	Sticky
*Bam*HI	*Bacillus amyloliquefaciens* H	5′- G|GATCC - 3′ 3′- CCTAG|G - 5′	Sticky
*Sal*I	*Streptomyces albus* G	5′- G|TCGAC - 3′ 3′- CAGCT|G - 5′	Sticky
*Bal*I	*Brevibacterium albidum*	5′- TGG|CCA - 3′ 3′- ACC|GGT - 5′	Blunt
*Hae*III	*Haemophilus aegyptius*	5′- GG|CC - 3′ 3′- CC|GG -5′	Blunt
*Sma*I	*Serratia marcescens*	5′- CCC|GGG - 3′ 3′- GGG|CCC - 5′	Blunt

a Using the table, explain the difference between *Hind*III and *Hae*III in the way they cut DNA? *1 mark*

b A researcher wanted to insert an insect-resistant gene into a plasmid. He accidently put DNA polymerase in with his plasmids instead of DNA ligase. Describe the difference between DNA polymerase and DNA ligase and describe the result of this error. *2 marks*

c The same researcher also cuts the genome containing the insect resistant gene with *Eco*RI because the recognition sites were conveniently located in the correct positions. He then used *Bam*HI to cut the plasmid because the recognition sequence was in the appropriate position on the plasmid to not interrupt any of the other genes. Would the researcher successfully create recombinant DNA? Explain. *2 marks*

Question 19 (6 marks)

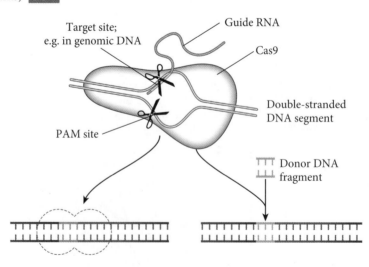

a Referring to the diagram above

  i describe the function of the guide RNA. *1 mark*

  ii describe the function of the PAM site. *2 marks*

b Outline the steps involved in the CRISPR process that lead to the insertion of a new gene. *3 marks*

Question 20 (2 marks) ●○○

CRISPR is a short sequence of DNA that is used as a target for CRISPR-associated nucleases (Cas nucleases). Researchers can use this technology to silence genes that produce proteins associated with allergy in peanuts. This will result in peanuts that do not cause an immune response in people. Discuss the usefulness of CRISPR technology in editing the genome of peanuts.   2 marks

Question 21 (6 marks) ●●○

The polymerase chain reaction can be used to identify the presence of a pathogen (a disease-causing microorganism) at an infection site.

a Describe the steps in the polymerase chain reaction.   4 marks

b How can this technology be used to identify the pathogen?   2 marks

Question 22 (2 marks) ●○○

a What is the name given to the DNA that consists of variable lengths of repeating DNA used in profiling?   1 mark

b How can this be used to identify specific individuals?   1 mark

Question 23 (5 marks) ●●○

A burglar broke into a jewellery store by smashing a glass window at the back of the building. When they were climbing in through the window, they cut themselves and left a small smear of blood on a shard of glass that was left in the frame. Three people were identified by neighbours as being seen in the area at the time of the burglary.

PCR was performed on DNA samples from all three suspects and then the samples were run through a gel electrophoresis. The results are shown at the right.

a What are the sizes of the fragments shown for Suspect 1?

b Whose blood was left on the glass shard and how do you know?

c Name **two** factors other than the size of the DNA fragment that can affect the rate at which the fragments migrate through the gel during electrophoresis.   2 marks

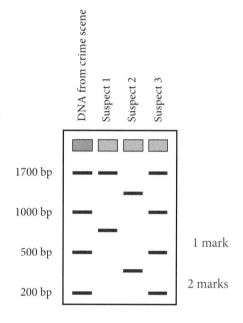

1 mark

2 marks

Question 24 (5 marks) ●●●

Researchers use recombinant DNA technology to insert desired genes into plasmids with the aid of restriction enzymes. These recombinant plasmids are then inserted into bacteria. One method to do this is a sharp change in temperature that opens pores in the bacterial membrane.

Step 1 – A tube containing a mixture of the enzymes, the desired gene and isolated bacterial plasmids are incubated on ice for 10 minutes.

Step 2 – The tube is then exposed to 42°C for exactly 50 seconds and then plunged back on ice.

a After step 1, would all plasmids contain the desired gene? Explain.   1 mark

b After step 2, will all bacteria accept the recombinant plasmids? Explain.   1 mark

c Describe an addition to this method that could be used by researchers to identify whether the bacteria have accepted a plasmid.   1 mark

d Explain how a protein product can then be produced in large quantities using the transformed bacteria and why bacteria are a good choice for this process.   2 marks

Question 25 (5 marks) ●●●

People with diabetes fail to make insulin or make a faulty insulin protein that does not work effectively. DNA technologies can be used to transform bacteria with an artificial insulin gene. This can then be used to produce insulin in large quantities.

Below are the steps required to create genetically engineered insulin. The steps have been mixed up.

A – Add solution with recombinant plasmids and bacteria quickly to a warm water bath to enhance the uptake of plasmids.

B – Look at the amino acid sequence and derive the nucleotide sequence from it.

C – Culture the transformed bacteria and then isolate the insulin produced.

D – Isolate the insulin protein from human cells and determine its sequence of amino acids.

E – Make recombinant plasmids – insert the artificial insulin gene into cut plasmids.

F – Identify the transformed bacteria that have the recombinant plasmid.

G – Make the artificial insulin gene in a gene synthesiser.

a   Why is it important to be able to produce insulin in commercial quantities?   1 mark

b   What is a plasmid and why can it be called a vector?   2 marks

c   What is the correct sequence of steps when producing the insulin from the jumbled sequence above?   1 mark

d   What is the tool or enzyme used for joining the artificial gene to plasmid DNA at step E?   1 mark

Question 26 (8 marks) ©VCAA VCAA 2018 SB Q10 ●●●

SHOULD WE GROW GM CROPS?
By Mary Nguyen

More than 25 years after genetically modified (GM) food first appeared, growing GM crops remains a hotly debated topic. Some people argue that GM crops are the only way to feed the growing world population and to minimise environmental harm. Other people express different views. Bt cotton is a type of cotton that contains two genes from a soil bacterium, *Bacillus thuringiensis*, enabling it to produce insect-resistant proteins. Australian farmers of Bt cotton use only 15% of the quantity of the insecticide that was once needed to protect their cotton crops[*]. However, Bt cotton is not as resistant to the main insect pest of cotton crops, *Helicoverpa*, as it has been in the past[*]. In Australia, Bt cotton is picked by machine, but in India, it is picked by hand. Workers in India have developed skin allergies, which have been attributed to Bt cotton proteins. Traditionally, farmers have saved money by keeping seed from one year's crop to plant the following year. However, it is illegal for farmers to keep Bt cotton seeds because these seeds have been declared the legal property of the company Monsanto. Every year, cotton farmers must buy more seeds from Monsanto. Unlike Monsanto, the company that produces the GM food crop Golden Rice allows farmers to replant the rice they harvested the previous year. By inserting a gene from the bacteria *Erwinia uredovora* and another from a daffodil, *Narcissus pseudonarcissus*, into white rice, scientists produced Golden Rice – a rice variety containing higher levels of vitamin A[†]. People who eat Golden Rice avoid vitamin A deficiency. Trials conducted in several countries have shown that Golden Rice is safe to eat[‡].

References: [*]CSIRO, 'Cotton pest management', case study;
[†]JA Paine et al., 'Improving the nutritional value of Golden Rice through increased pro-vitamin A content', *Nature Biotechnology*, 23, 27 March 2005, pp. 482–487;
[‡]A Coghlan, 'Golden Rice gets approval in the US', *New Scientist*, magazine issue 3180, 2 June 2018

a   Bt cotton and Golden Rice are genetically modified organisms but are they also transgenic organisms? Support your response with evidence from the article above.   3 marks

b   How can planting a Bt cotton crop lead to an increase in crop yield?   1 mark

c   Using information from the article, copy and complete the table below by describing one social
and one biological implication relevant to the use of Bt cotton and Golden Rice. The same
implication should not be used twice.   4 marks

	Social implication	Biological implication
Bt cotton		
Golden rice		

Question 27 (6 marks) ⬤⬤⬤

Soil salinity is a major problem that affects crop productivity in many areas of the world. In areas where
there is very little rainfall, farmers are forced to use salty water to irrigate their crops. Researchers have
discovered a protein called BspA that is common in the tree *Populus tremula*. This protein enables this
plant to tolerate high salinity due to its ability to attract water molecules.

Researchers found that increasing the number of genes that code for this protein within plant cells
resulted in increased salt tolerance in tomato plants. The DNA was transferred into the plant cells and
these were then grown in a laboratory. Once these cells develop into a plant, the seeds produced from this
plant will inherit the new DNA.

a   Suggest a method that could be used to insert the genes into plant cells.   1 mark

b   Outline an experimental design you could use to determine whether the seeds harvested from the
transformed plants have a higher salt tolerance to other wild type tomato plants. Remember to include
all relevant components in sound experimental design.   5 marks

Chapter 2 Area of Study 2
How are biochemical pathways regulated?

Area of Study summary

This Area of Study focuses on the function of enzymes and coenzymes and the environmental factors that influence them. Energy is stored in chemical bonds that can be released through biochemical pathways. Enzymes and coenzymes regulate biochemical pathways such as photosynthesis and cellular respiration through a series of steps that allow the organisms to adapt to changing environments, and ensure their energy requirements meet their activity. Knowledge of enzymes and coenzymes and the factors that affect them can be used in applications of biotechnology to benefit industry, medicine and agriculture.

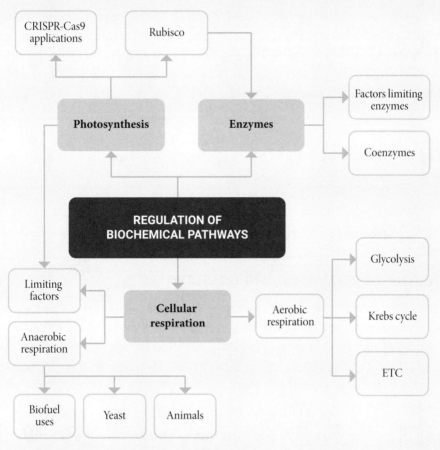

Area of Study 2 Outcome 2

On completing this outcome, you should be able to:

- **analyse** the structure and regulation of biochemical pathways in photosynthesis and cellular respiration
- **evaluate** how biotechnology can be used to solve problems related to the regulation of biochemical pathways.

The key science skills demonstrated in this outcome are:

- **analyse**, **record** and **collate** data
- **evaluate** data and investigation methods.

2.1 Regulation of biochemical pathways in photosynthesis and cellular respiration

Photosynthesis and **cellular respiration** involve molecular transformations that occur within specific areas of cells and require multiple steps from initial reactants to final products. Biochemical pathways are a series of reactions that can involve the use of enzymes and coenzymes to regulate each step. It is important for these reactions to be regulated so the cell can match its energy requirements to its activity. Photosynthesis and cellular respiration both use pathways that can be turned 'up' or 'down', using enzymes to control them.

2.1.1 General structure of biochemical pathways in photosynthesis and cellular respiration

The steps that form a biochemical pathway, also known as a metabolic pathway, involve a series of enzyme-mediated reactions, where the product of one reaction is used as the substrate for the next reaction. This allows the cell to respond to changing environmental demands to regulate these pathways, which are essential for survival. The enzymes are able to respond through metabolic signals. If one step of a pathway is active, the next step can take place quickly. However, if the enzyme is inactive, the pathway will not progress.

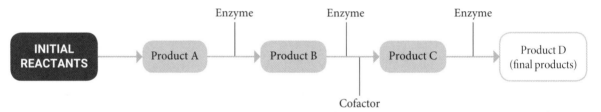

A biochemical pathway where the products of one stage are the substrates for the next stage

2.1.2 The role of enzymes and coenzymes

Enzymes are first described in Unit 3, Area of Study 1. You will recall that they are proteins that catalyse biological reactions by lowering the activation energy required for a reaction to proceed.

Enzymes are highly specific in nature due to the shape of their active site. They can be involved in catalysing each step of the biochemical pathway. At each step, the enzyme helps to transform a molecule into another form so that at the end of the pathway a new molecule is synthesised or broken down.

Coenzymes are non-protein organic molecules that cannot catalyse reactions by themselves. However, because they sit in the active site of enzymes, they are necessary to assist in the enzyme-controlled reactions of photosynthesis and respiration. A coenzyme is a type of cofactor that carries electrons and protons, or hydrogen ions, between enzymes.

There are four coenzymes involved in photosynthesis and cellular respiration. Coenzymes are non-protein organic molecules that cycle between stages to shuttle electrons, protons and hydrogen ions to ensure that each step of the biochemical pathway can occur.

Note
Refer to page 22 to refresh your knowledge on enzymes.

H⁺ pump in electron transport system

Energy from electron transfers

ATP synthase

$ADP + P_i \rightarrow ATP$

A gradient set up across a membrane where H⁺ will diffuse from an area of high concentration to an area of low concentration through ATP-synthase. This flow of protons provides the energy required to synthesise ATP from ADP + P_i.

In both photosynthesis and cellular respiration, the transfer of electrons sets up an **electrochemical gradient**, allowing **chemiosmosis** to occur. The gradient aids hydrogen ions, or protons, to move through a protein channel called **ATP-synthase**, which synthesises ATP from ADP and Pi (inorganic phosphate). This process is called **oxidative phosphorylation**.

The roles and forms of the four coenzymes are outlined below.

Coenzyme loaded	Unloaded form	Role of coenzyme	Reaction
NADH	NAD^+	Transfer of electrons and protons in cellular respiration	$NADP^+ + 2e^- + 2H^+ \rightarrow NADPH + H^+$
NADPH	$NADP^+$	Transfer of electrons and proteins in photosynthesis	$NADP^+ + 2e^- + 2H^+ \rightarrow NADPH + H^+$
ATP	ADP	Energy transfer in cellular respiration and photosynthesis	$ADP + P_i \rightarrow ATP$
$FADH_2$	FAD	Transfer of electrons and proteins in cellular respiration	$FAD + 2e^- + 2H^+ \rightarrow FADH_2$

2.1.3 General factors that impact on enzyme function

Enzymes function best when they are in their optimum environment. When the environment fluctuates outside of the enzyme's optimum range it will impact their function and may even cause denaturation of the enzyme. Denaturation is the process where the molecular structure of the enzyme is compromised. The bonds that hold the tertiary structure of the enzyme together are broken and cannot be reformed. This will result in the enzyme no longer being able to function: the active site of the enzyme has changed shape, so enzyme–substrate complexes cannot form and reactions cannot be catalysed.

The function of an enzyme can be measured:

1 by the concentration of reactant or substrate used up in a given time

2 by the concentration of product formed in a given time.

The factors impacting enzyme function in photosynthesis and respiration and their impacts on the biochemical pathways are outlined below.

Factor	Graph	Explanation
pH		The optimum pH is where the rate of reaction is highest. As the pH level deviates from the optimum, the rate of reaction decreases to zero as the enzyme becomes denatured.
Temperature		At temperatures cooler than the enzyme's optimum, there is less kinetic energy so the rate of reaction will be slow as it takes longer to form enzyme substrate complexes. Maximum rate of reaction occurs at the optimum temperature. At temperatures higher than the enzyme's optimum, the enzyme denatures and no reaction can occur.

》

Factor	Graph	Explanation
Substrate concentration	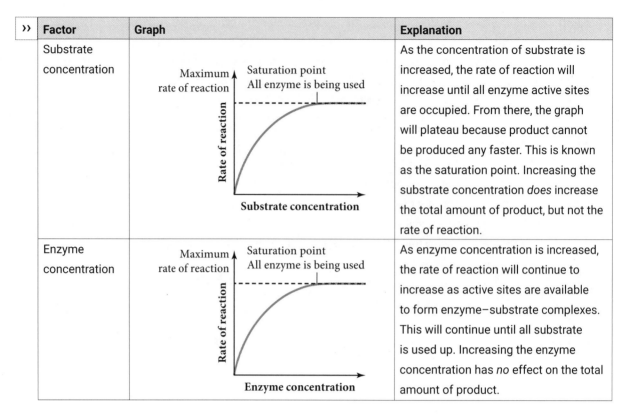	As the concentration of substrate is increased, the rate of reaction will increase until all enzyme active sites are occupied. From there, the graph will plateau because product cannot be produced any faster. This is known as the saturation point. Increasing the substrate concentration *does* increase the total amount of product, but not the rate of reaction.
Enzyme concentration		As enzyme concentration is increased, the rate of reaction will continue to increase as active sites are available to form enzyme–substrate complexes. This will continue until all substrate is used up. Increasing the enzyme concentration has *no* effect on the total amount of product.

Enzyme models

The **lock-and-key model** of enzyme action proposes that the shape of the enzyme's active site is highly specific for a particular substrate. The substrate's shape is complementary to the shape of the active site within the enzyme so it fits into it like a key fits into a lock.

The **induced-fit model** of enzyme action proposes that the bonds that form between the enzyme and its substrate modify the shape of the active site so that the substrate can fit snugly. To achieve this the bonds of the substrate are stretched and bent. The new product molecules are not specific to the active site and are released.

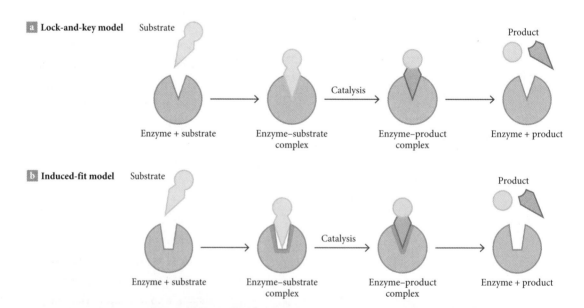

Enzyme action. a In the lock-and-key model, the binding of the substrate into the active site of an enzyme mirrors a door's lock-and-key mechanism. The substrate's shape is complementary to the shape of the active site within the enzyme. b In the induced-fit model, the substrate molecule enters the enzyme's active site, causing the enzyme molecule to change shape so that the two molecules fit together more closely.

9780170479431

How inhibitors impact the function of enzymes

Enzyme action can be affected by the presence of other molecules that may inhibit their action. Some poisons and antibiotics act as inhibitors; for example, herbicides prevent plant enzymes from carrying out photosynthesis, resulting in the death of the plant. Similarly, cyanide is an irreversible inhibitor that inhibits cytochrome c oxidase during cellular respiration, resulting in an organism's death.

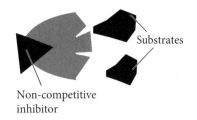

A non-competitive inhibitor binds to the enzyme at the allosteric site, which causes a conformational change of the active site, preventing the formation of an enzyme–substrate complex.

Inhibition of enzymes can be either reversible or irreversible and their mode of action can be either competitive or non-competitive inhibition. Irreversible inhibition is due to strong covalent bonds, which are difficult to break. Reversible inhibition has weaker bonds that are much easier to break.

Competitive inhibition

A **competitive inhibitor** is a molecule that competes with the substrate for binding to the enzyme's active site. It binds temporarily with the active site, preventing the binding of enzyme and substrate.

Non-competitive inhibition

A **non-competitive inhibitor** is a molecule that binds with the enzyme in another part of the molecule called the allosteric site. This is not at the active site. These inhibitors alter the shape of the enzyme molecule and therefore affect the ability of the active site to bind with substrate molecules.

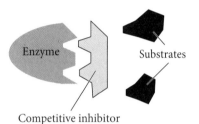

A competitive inhibitor competes with a substrate for the active site.

2.2 Photosynthesis as an example of biochemical pathways

Photosynthesis is a biochemical pathway that takes place in the chloroplasts in plants whereby carbon dioxide in the atmosphere is fixed in order to produce glucose for the plant. It occurs in two main stages, where the outputs of one stage are the inputs for the second stage. Photosynthesis requires an input of energy to form the products; this makes it an anabolic and endergonic reaction. The coenzymes required cycle between the stages, regulating the reactions and providing the energy required to help fuel them.

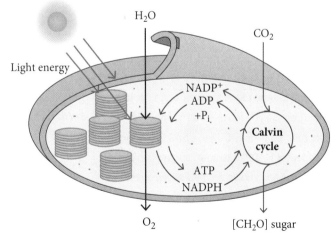

A summary of two stages of photosynthesis, including the inputs and outputs of these stages and the cycling of coenzymes between them

2.2.1 Inputs, outputs and locations of the light-dependent and light-independent stages of photosynthesis

Spinach, peanuts, cotton, rice, wheat and most trees and grasses are all examples of C_3 plants. Like all plants, they rely on photosynthesis to produce glucose, which is essential for the plant because it is the substrate for cellular respiration. The reactions that make up the process of photosynthesis can be divided into two stages: **light-dependent stage** and **light-independent stage**. (Light-independent reactions are sometimes called dark reactions. Despite their name, they do not require darkness; they simply don't need light.)

Photosynthesis occurs in the chloroplast. These organelles are green in colour because they contain the pigment chlorophyll. It is the pigments contained within the chloroplast that are used to absorb sunlight for energy. As chlorophyll and light are essential for photosynthesis because they provide the energy, they are written above and below the arrow in the reaction. They cannot be written on the left, as only reactants are included there. They cannot be written on the right, as only the products of the reaction are written here.

The overall reaction for photosynthesis can be written as the below equation:

$$\text{Inputs} \qquad\qquad\qquad\qquad \text{Outputs}$$
$$[\text{Carbon dioxide} + \text{NADPH} + \text{ATP}] \xrightarrow[\text{Light}]{\text{Chlorophyll}} [\text{glucose} + \text{water} + \text{NADP}^+ + \text{ADP} + \text{P}_i]$$

The steps are summarised in the table and diagram below.

Step	Input	Location	Products
Light-dependent stage Light energy is trapped by chlorophyll and used to split water molecules into H^+ ions and oxygen. The oxygen is given off as waste. The energy produced from splitting water is used to create ATP. The H^+ ions are taken up by acceptor molecules of NADP. $12H_2O + \text{light energy} \longrightarrow 6O_2 + 24H^+$ (as NADPH) + stored chemical energy as ATP	Light Water ADP + P_i $NADP^+$	Within grana (thylakoid membranes)	$6O_2$ (from the split water molecules; it is released to the atmosphere) ATP NADPH (these loaded coenzymes are used in the light-independent reactions)
Light-independent stage (Calvin cycle) Carbon dioxide from the atmosphere is combined with hydrogen (from the loaded acceptor NADPH) to form sugars in a series of reactions known as the **Calvin cycle**. The ATP produced in the light reactions provides the energy needed for this reaction. $6CO_2 + 24H^+$ (from NADPH) + energy (from ATP) $\longrightarrow C_6H_{12}O_6 + 6H_2O$	Carbon dioxide NADPH ATP	In stroma	Glucose $C_6H_{12}O_6$ $6H_2O$ $NADP^+$ ADP

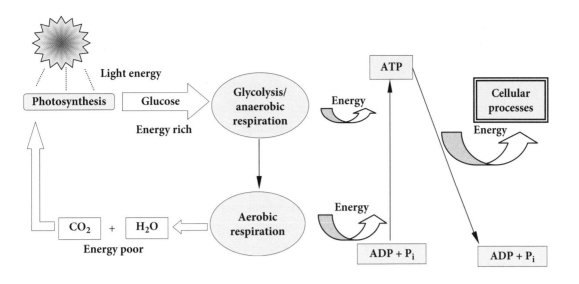

2.2.2 The role of Rubisco in photosynthesis

The enzyme **Rubisco**, short for ribulose-1,5-bisphosphate carboxylase/oxygenase, catalyses the fixation of inorganic, atmospheric carbon in the form of carbon dioxide, into an organic carbon molecule. The difference between organic and inorganic carbon is that in an organic carbon molecule there is at least one carbon bonded to a hydrogen atom, forming a C–H bond. So, when CO_2 is fixed, it forms a new molecule.

In C_3 plants, which we mentioned in 2.2.1, Rubisco fixes CO_2 straight from the atmosphere into a 3-carbon organic compound and eventually, through the Calvin cycle, glucose is produced.

Rubisco can also catalyse a reaction where oxygen present in the atmosphere is converted into a product that is not a sugar compound through steps in the Calvin cycle. This is known as **photorespiration** and it is a wasteful, inefficient pathway. At low temperatures CO_2 is more likely to bind to Rubisco than O_2.

As oxygen is a competitive inhibitor for Rubisco, C_3 plants have reduced rates of photosynthesis when oxygen is present. In order to minimise this, C_3 plants can keep their stomata open to allow oxygen produced during the light-independent stage of photosynthesis to diffuse out of the leaf, keeping CO_2 concentrations higher than O_2 concentrations.

This is very difficult in hot and dry climates because stomata must remain closed to prevent water loss. Other plants, known as C_4 and CAM plants, have evolved to limit the exposure of Rubisco to oxygen.

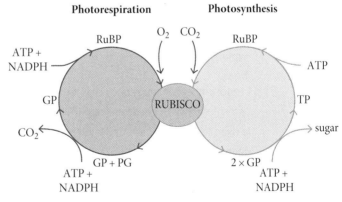

> **Note**
> C_4 and CAM plants are plants that have adaptations to use special compounds to gather carbon dioxide during photosynthesis. This allows these plants to extract more CO_2 from the atmosphere, while preventing water loss in warmer climates.
>
> ››

Pathway	Method	Mode of action	Climate and examples
C_3	CO_2 completes with oxygen to bind to Rubisco At low temperatures the rate of photorespiration is low and photosynthesis is high. This reverses as temperatures increase.	CO_2 is converted into a C_3 product by Rubisco.	Best adapted to cool, wet environments Examples: plants, wheat, rye, rice cotton

Pathway	Method	Mode of action	Climate and examples
C_4	Carbon dioxide is physically separated from oxygen in order to improve CO_2 binding to Rubisco.	CO_2 is converted to a C_4 compound using a different enzyme. It is then moved deeper into the tissue layer of the plant where less oxygen is present. Here, the C_4 product is broken down and CO_2 is released which can enter the Calvin cycle and be fixed by Rubisco without competition from oxygen.	Best adapted to hot, sunny, tropical environments Examples: maize, sugar cane
CAM	Carbon dioxide reserves are formed in order to improve CO_2 binding to Rubisco.	CO_2 is converted into a C_4 compound by a different enzyme during the night, when stomata are open and CO_2 is able to diffuse into the leaf. The C_4 compounds are stored for breakdown during the day, where they will be broken down and CO_2 is released. The stomata are closed and oxygen cannot be released but the concentration of CO_2 should be higher so it can bind to Rubisco to form glucose through the steps in the Calvin cycle.	Very hot, dry environments Examples: cacti and agave.

2.2.3 Factors that affect the rate of photosynthesis

Factors that affect the rate at which photosynthesis occurs are called 'limiting factors' because they limit how fast product can be made. If the rate of photosynthesis decreases, the production of glucose also decreases. This means less chemical energy is available for the cell and growth will be compromised. There are a number of limiting factors that will impact the rate of photosynthesis on their own, regardless of the level of other factors. The major limiting factors are outlined below.

Limiting factor	Explanation	Graph
Light availability	Photosynthesis cannot occur without light. Plants use chlorophyll to absorb sunlight. The light energy is used to split the water molecules in the light-dependent stage. As more sunlight is absorbed, particularly in the blue and red ranges, the rate of photosynthesis increases. At a certain light level, the rate tends to stop increasing due to a different limiting factor, such as CO_2. Water or chloroplast enzyme active sites are all occupied. At extremely high light levels, the rate might even decrease because the chlorophyll molecules can become damaged. This level varies from plant to plant, which is why some plants need to be kept in the shade, whereas others do perfectly well in direct sunlight.	

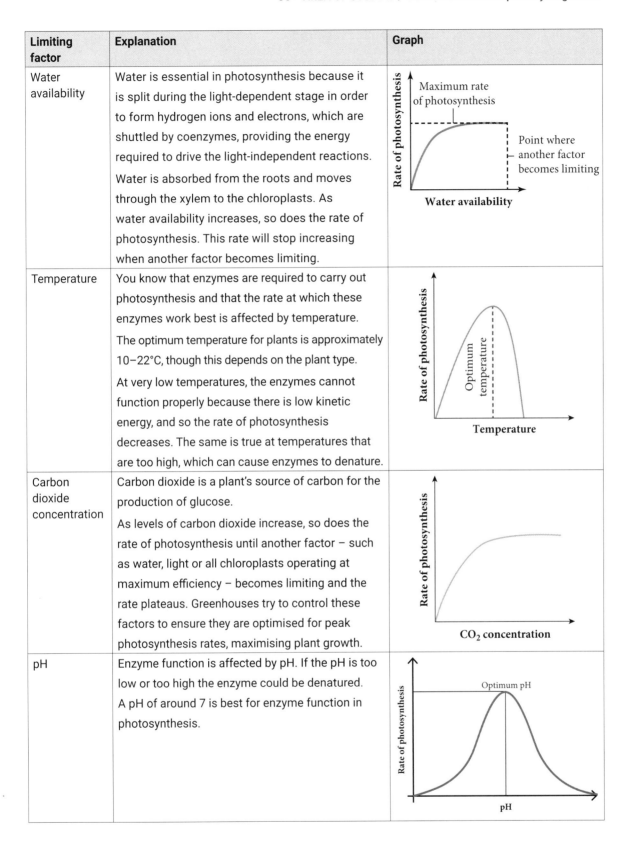

Limiting factor	Explanation	Graph
Water availability	Water is essential in photosynthesis because it is split during the light-dependent stage in order to form hydrogen ions and electrons, which are shuttled by coenzymes, providing the energy required to drive the light-independent reactions. Water is absorbed from the roots and moves through the xylem to the chloroplasts. As water availability increases, so does the rate of photosynthesis. This rate will stop increasing when another factor becomes limiting.	
Temperature	You know that enzymes are required to carry out photosynthesis and that the rate at which these enzymes work best is affected by temperature. The optimum temperature for plants is approximately 10–22°C, though this depends on the plant type. At very low temperatures, the enzymes cannot function properly because there is low kinetic energy, and so the rate of photosynthesis decreases. The same is true at temperatures that are too high, which can cause enzymes to denature.	
Carbon dioxide concentration	Carbon dioxide is a plant's source of carbon for the production of glucose. As levels of carbon dioxide increase, so does the rate of photosynthesis until another factor – such as water, light or all chloroplasts operating at maximum efficiency – becomes limiting and the rate plateaus. Greenhouses try to control these factors to ensure they are optimised for peak photosynthesis rates, maximising plant growth.	
pH	Enzyme function is affected by pH. If the pH is too low or too high the enzyme could be denatured. A pH of around 7 is best for enzyme function in photosynthesis.	

2.3 Cellular respiration as an example of biochemical pathways

Glucose ($C_6H_{12}O_6$) is an energy-rich molecule that can be used by cells as an energy source. Glucose is broken down to release energy in a catabolic/exergonic reaction. In the presence of oxygen, most cells carry out cellular respiration or **aerobic respiration**. Aerobic respiration can be summarised as follows:

$$C_6H_{12}O_6 + 6\,O_2 \longrightarrow 6\,CO_2 + H_2O + 30 \text{ or } 32 \text{ ATP}$$

This reaction is another example of a biochemical pathway that uses enzymes and coenzymes in a series of steps that can be regulated because the outputs of one stage are the inputs for another.

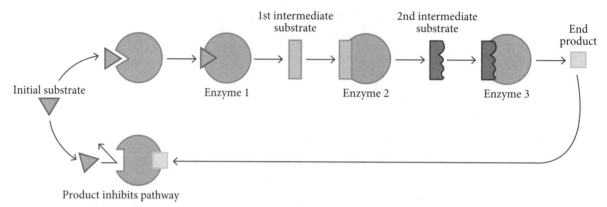

Cellular respiration as an example of a biochemical pathway regulated by steps where the product of each stage is required as an input for the next stage to occur

2.3.1 Main inputs, outputs and locations of glycolysis, Krebs cycle and electron transport chain

The three steps involved in cellular respiration are outlined below, along with the location where each step occurs and the inputs and outputs of each.

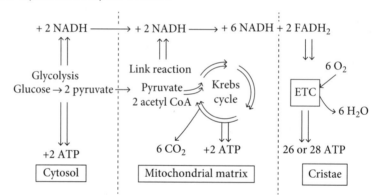

Step	Location	Inputs	Outputs
Glycolysis In **glycolysis**, glucose (a six-carbon molecule) is split into two pyruvate (a three-carbon molecule) molecules. Energy is released. Energy is used to form ATP and the loaded acceptor molecule NADH.	In the cytosol of cells	1 glucose 2 ADP + P_i 2 NAD$^+$	2 pyruvate 2 ATP 2 NADH molecules that can be used later to produce more ATP
Krebs cycle In the **Krebs cycle**, each pyruvate molecule is broken down and three molecules of carbon dioxide are formed. Some ATP and more loaded acceptors are formed.	Matrix – the inner compartment of mitochondria	2 pyruvate converted to 2 acetyl CoA 2 ADP + P_i 8 NAD$^+$ 2 FAD$^+$	2 ATP 6 CO_2 8 NADH 2 FADH$_2$

Step	Location	Inputs	Outputs
Electron transport chain Electrons from the loaded molecules NADH and $FADH_2$ are transferred along the chain to be finally accepted by oxygen. This is known as the **electron transport chain**. The oxygen reacts with hydrogen to form water. Energy is released.	Cristae – the inner membranes of mitochondria	6 oxygen 26 or 28 ADP + P_i 2 $FADH_2$	26 or 28 ATP in most cells 10 NAD^+ 2 FAD^+ 6 H_2O

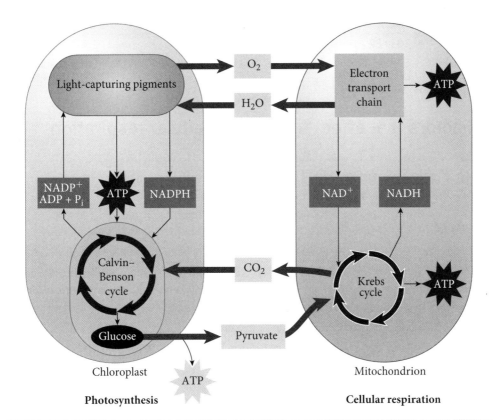

The inter-relationship between photosynthesis and cellular respiration

2.3.2 Location, inputs and the difference in outputs of anaerobic fermentation in animals and yeasts

Photosynthesis, aerobic respiration, anaerobic respiration and anaerobic fermentation are processes used to extract energy from molecules. The major difference between anaerobic fermentation and aerobic respiration is that anaerobic fermentation is much less efficient than aerobic respiration but it will occur in the absence of oxygen.

Anaerobic respiration occurs in animals in the cytosol of cells. It will occur when there is not enough energy being produced by aerobic respiration, such as during strenuous exercise. Cells simply can't get oxygen fast enough to produce all the energy they need. Anaerobic respiration produces ATP more rapidly than aerobic respiration does; however, it cannot be undertaken for any length of time because lactic acid can build up in the body's muscles, causing pain and fatigue.

$$C_6H_{12}O_6 \rightarrow 2 \text{ lactic acid} + 2 \text{ ATP}$$

Anaerobic respiration in yeasts is known as anaerobic fermentation. The process of anaerobic fermentation has many commercial uses, such as baking and beer and wine production. In wine and beer, sugars are metabolised to form ethanol, or drinking alcohol. In baking, this same process causes bread to rise with the production of CO_2. During the baking process, most of the ethanol evaporates into the atmosphere.

$$Glucose \longrightarrow ethanol + carbon\ dioxide + ATP$$

$$C_6H_{12}O_6 \longrightarrow 2\ C_2H_5OH + 2\ CO_2 + 2\ ATP$$

	Aerobic respiration	Anaerobic respiration	Anaerobic fermentation
Oxygen required	Yes	No	No
ATP yield	30 or 32 ATP	2 ATP	2 ATP
End products	Carbon dioxide and water	Lactic acid	Ethanol and carbon dioxide
Location	Cytosol and mitochondria	Cytosol	Cytosol

2.3.3 Factors that affect the rate of cellular respiration

Cellular respiration requires optimum conditions to be able to continually produce the ATP that is required by cells to function. Some factors that affect how quickly this process can occur, and can impact how much ATP can be produced, are explained below.

Temperature

Cellular respiration requires enzymes to catalyse reactions. We already know that enzymes have an optimum temperature, and the enzymes involved in metabolic processes in the human body have an optimum temperature of 37 degrees Celsius. This is why, on a cold day, your body will trigger a response for you to shiver, generating heat and warming you up. Conversely, on a hot day, you will sweat as a way for your body to lose heat to your surroundings.

If for any reason your body temperature is too low, the enzymes slow down and the rate of reaction decreases.

If your body temperature rises too high, the enzymes start to denature. The bonds holding together the 3D structure can be broken and the enzymes are not able to catalyse reactions anymore because they cannot form enzyme–substrate complexes. For maximum efficiency, the temperature needs to be within the optimum range. This is why having a fever can be so dangerous.

Glucose availability

Glucose is produced through the digestion process, where carbohydrates are broken down into their smallest subunits and absorbed into the blood. Glucose travels in the blood to respiring cells, where it is broken down through glycolysis in the first stage of cellular respiration. This stage is essential for both aerobic and anaerobic respiration and, if there is not enough glucose available, the cell will not be able to produce ATP. As the availability of glucose increases, the rate of cellular respiration increases until another factor becomes limiting.

Oxygen concentration

Oxygen enters the body through ventilation, when air enters our lungs and oxygen crosses into the bloodstream through diffusion. Oxygen is essential for the final stage of cellular respiration: the electron transport chain. It acts as the final electron acceptor, able to accept two hydrogen ions and two electrons, forming water. This allows for the continuous flow of electrons through the chain. If oxygen is not present, the electrons will not pass through the chain and ATP cannot be produced, resulting in the death of cells. As oxygen concentration is increased, the rate of aerobic cellular respiration increases until another factor becomes limiting. Oxygen is not an input of anaerobic respiration and thus would not be a limiting factor, having no influence on the rate of reaction.

2.4 Biotechnological applications of biochemical pathways

The more scientists know and understand about biochemical pathways and the transformation of energy, the more these processes can be manipulated through biotechnological applications. These applications can lead to advancements in many areas.

With population growth and climate change increasing the demand on crop productivity, there is an increasing need to overcome a potential crisis of food shortage. Plants require photosynthesis to build **biomass** so, by increasing the efficiency and productivity of photosynthesis in crops, scientists are able to improve crop yields.

Earlier in this unit you learnt of the factors that impact the rate of photosynthesis and also the different pathways that C_3, C_4 and CAM plants are able to use to fix carbon. In particular, you learnt how C_4 and CAM plants have evolved to reduce the inefficient pathway of photorespiration, where oxygen competes with CO_2 for the enzyme Rubisco. Using CRISPR-Cas9 technology, DNA sequences in plant genomes can be modified to enhance photosynthetic pathways by altering plant structure, increasing light capture and energy conservation. CRISPR can target the exact **genes** or regulatory genes far more cheaply and quickly than cross-breeding or transgenic methods.

Targets of opportunity	Applications
Light capture and energy conversion	Currently, plants absorb more light than they can use. Modifying light-harvesting pigments, such as chlorophyll, to reduce their quantity in a plant's uppermost leaves would allow plants to absorb less light but increase their efficiency, leading to better conversion of light to biomass through the reduction the plant's energy spent on diverting excess light. This application is currently seen in algae and cyanobacteria.
Carbon capture and conversion	Using CRISPR to introduce CO_2 chambers into photosynthetic cells would allow for higher CO_2 concentration in cells and increase the chance that CO_2 will bond to Rubisco, outcompeting oxygen. These channels are utilised in algae and cyanobacteria. Modified Rubisco could reduce the impact of climate change by creating crops that thrive with increased carbon dioxide levels and higher temperatures. Engineering rice and maize to use the C_4 pathway could lead to them being more efficient in hot and dry environments.

CRISPR can also be used in other potential areas such as improving virus and pest resistance, improving starch yield in potato crops, and introducing novel genes into plants to enhance their efficiency, such as improving fibre quality in cotton production.

2.4.1 Uses and applications of anaerobic fermentation

Biofuel is any fuel that comes from the breakdown of biomass (algal, plant or animal waste). Since biomass is readily available and easily replenished, biofuel is considered to be a source of sustainable energy, unlike fossil fuels, which include coal, petroleum and natural gas. Furthermore, biofuels do not cause a net increase to atmospheric greenhouse gases because the carbon source is maintained in a closed system where carbon dioxide in the atmosphere is drawn into the plant through photosynthesis. Livestock then eat the plants and excrete excess carbon as manure. Microorganisms break down organic matter such as manure, sewage and food scraps to produce the methane as a by-product. The production of biofuel, or methane, can be controlled in fermenters. Fermenters must be airtight so no oxygen can enter, and must maintain a pH of 7 and a temperature of 37°C. They must have an inlet for the organic matter and contain a vessel to trap the methane, which can be burned to produce CO_2 and water vapour, leaving no carbon footprint.

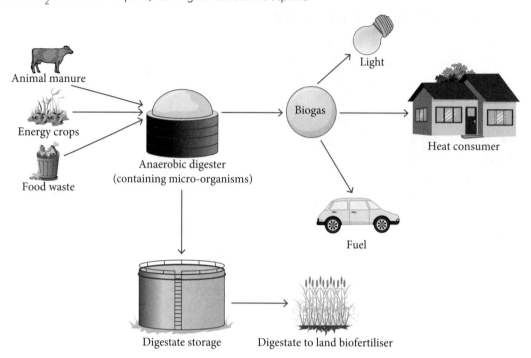

Organic material is broken down by micro-organisms to produce biogas, which can be used for electricity and fuel.

Uses for biofuel	Technology
Transport	Biofuel can be turned into hydrogen steam that can be used in fuel cells in powering electric vehicles.
Electricity generation	Biofuel can be used to generate power back-up systems in schools, hospitals and residential areas. The electricity is generated through the burning of methane.
Heat production	Methane gas is burned to provide a heat source.

Glossary

aerobic respiration An energy-releasing reaction that occurs within cells; it requires oxygen and glucose in order to produce chemical energy, ATP; occurs in the mitochondria

anaerobic respiration An energy-releasing reaction that occurs within cells; it does not require oxygen; occurs in the cytosol; plants produce carbon dioxide, ethanol and ATP and humans produce lactic acid and ATP

ATP-synthase An enzyme that catalyses the formation of ATP from ADP + P_i using the potential energy stored in the hydrogen ion gradient created through chemiosmosis

biofuel A fuel that has used biomass as its original source

biomass The total dry weight of organic material

Calvin cycle A biochemical pathway in which sugar molecules are produced from carbon dioxide

cellular respiration The process in which cells convert glucose into ATP; can occur aerobically or anaerobically

chemiosmosis The movement of ions, such as H^+ ions, across a membrane down an electrochemical gradient

coenzyme An organic molecule; most coenzymes are carrier molecules that transfer electrons or ions from one reactant to another in a biological reaction

competitive inhibitor A molecule that competes with a substrate by binding temporarily to an enzyme's active site

electrochemical gradient The difference in charge and chemical concentration where ions will move across a membrane from an area of high concentration to an area of lower concentration

electron transport chain The stepwise transport of electrons to a final electron acceptor, such as oxygen (in aerobic cellular respiration); ultimately, it creates an electrochemical gradient across membranes to drive the phosphorylation of ADP to yield ATP

gene The basic unit of heredity

glycolysis The first step in aerobic and anaerobic respiration; involves the splitting of glucose (a six-carbon molecule) into two three-carbon molecules known as pyruvate

A+ DIGITAL FLASHCARDS
Revise this topic's key terms and concepts by scanning the QR code or typing the URL into your browser.

https://get.ga/a-biology-vce-u34

induced fit model Model of enzyme function that proposes that the enzyme changes shape to accomodate the substrate

Krebs cycle A biochemical pathway that requires oxygen and takes place in the mitochondria as part of cellular respiration; acetyl CoA, the product of glycolysis, is broken down to produce carbon dioxide, water and energy in the form of ATP

light-dependent stage The first stage of photosynthesis; it requires light energy that is absorbed by chlorophyll to split water molecules to produce oxygen, hydrogen ions and ATP

light-independent stage The second stage of photosynthesis; through a series of reactions carbon dioxide, hydrogen ions and ATP produce carbohydrates

lock and key model Model of enzyme function that proposes that the active site of the enzyme is an exact geometric fit for one specific substrate

non-competitive inhibitor A molecule that binds temporarily with an enzyme in a part of the molecule other than the active site; it inhibits enzyme action by altering the shape of the molecule

oxidative phosphorylation The process of transferring electrons from coenzymes via the electron transfer chain in the mitochondria, which results in the formation of ATP

photorespiration A process in plants where the enzyme Rubisco acts on oxygen, which is wasteful for the plant

photosynthesis The conversion of light energy to stored chemical energy in glucose molecules

Rubisco An enzyme involved in converting atmospheric carbon dioxide into carbon, which can be used by a plant to generate glucose

Revision summary

In the table below, provide a brief definition of each of the key concepts. This will ensure that you have revised all the key knowledge in this Area of Study in preparation for the exam.

How are biochemical pathways regulated?	
The overall structure of biochemical pathways from reactants to products for photosynthesis	
The overall structure of biochemical pathways from reactants to products for cellular respiration	
The role of enzymes and coenzymes in photosynthesis	
The role of enzymes and coenzymes in cellular respiration	
Models of enzyme function	
Factors that impact enzyme activity in biochemical pathways	
Inputs, outputs and locations of the two stages of photosynthesis in C_3 plants	
The role of Rubisco in adaptations of C_3, C_4 and CAM plants	
Limiting factors of photosynthesis	
The inputs, outputs and locations of glycolysis, Krebs cycle and the electron transport chain	
The locations, inputs and difference in outputs of anaerobic fermentation in animals and yeast	
Limiting factors of cellular respiration	
Possible uses and application of CRISPR-Cas9 technology to improve photosynthetic efficiency	
Uses and applications of anaerobic fermention of biomass for biofuel production	

Exam practice

Regulation of biochemical pathways in photosynthesis and cellular respiration

Solutions start on page 222.

Multiple-choice questions

Question 1 ◉○○

A biochemical pathway is a sequence of reactions. Which of the following is NOT true of a biochemical pathway?

A The enzymes catalysing the reactions are critically important in the progression of the pathway.

B The products of one reaction are the reactants of the next.

C Photosynthesis and cellular respiration are both examples of biochemical pathways.

D A biochemical pathway is a series of reactions that happen alongside one another and only interact at the very end, where the products combine to make the functional unit.

Question 2 ©VCAA VCAA 2017 SA Q5 ◉◉◉

The biochemical pathway of glycolysis involves nine intermediate reaction steps. One of these steps is represented in the diagram below.

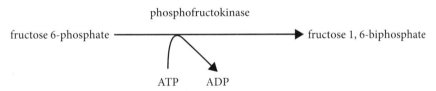

phosphofructokinase

fructose 6-phosphate ──────────────► fructose 1, 6-biphosphate

ATP   ADP

It is correct to state that, in this reaction, phosphofructokinase

A acts as a coenzyme.

B increases the rate of reaction.

C is the substrate for the reaction.

D releases energy in the form of ADP.

Question 3 ◉○○

Coenzymes are

A non-protein organic molecules that cannot catalyse reactions by themselves, but are necessary for enzyme-catalysed reactions to proceed.

B two enzymes that both catalyse the same reaction to achieve a high level of products at a fast rate.

C proteins that help reactions to proceed by labelling and transporting the reactants, thus reducing the energy required for the enzyme to catalyse the reactants into a product.

D organic molecules that can also catalyse reactions, similar to protein enzymes.

Question 4 ◉○○

Some of the coenzymes associated with photosynthesis and respiration include

A $NADH$, CTP, $FADH_2$.

B $FADH_2$, ATP, NAP.

C ATP, $NADH$, $NADPH$.

D $FADH_2$, ATP, NAH.

Question 5 ©VCAA VCAA 2014 SA Q13 ●●

The following graphs show the way four enzymes, W, X, Y and Z, change their activity in different pH and temperature situations.

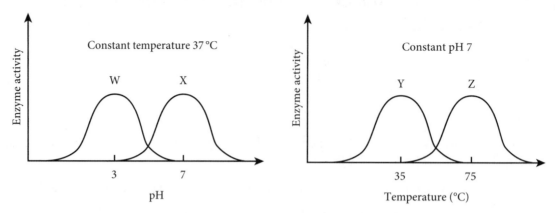

Which one of the following statements about the activity of the four enzymes is true?

A At pH 7, enzyme Y is denatured at temperatures below 20°C.

B Enzyme Z could be an intracellular human enzyme.

C At pH 3 and a temperature of 37°C, the active site of enzyme W binds well with its substrate.

D At pH 3 and a temperature of 37°C, enzyme X functions at its optimum level.

Question 6 ©VCAA VCAA 2018 SA Q7 ●●

Four students performed a series of experiments to investigate the effect of four different variables on the rate of an enzyme-catalysed reaction. In each experiment the students changed one of the following variables: substrate concentration, pH, temperature and enzyme concentration. After recording their data, the students displayed their results in a series of graphs, as shown below. Each graph is a line of best fit for their data.

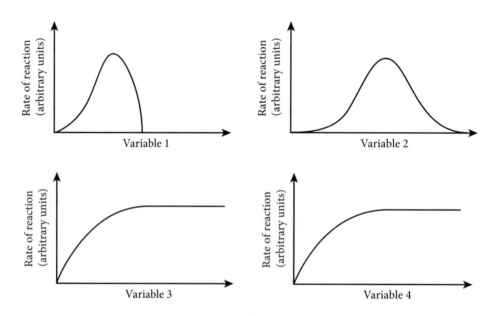

The students did not label the horizontal axis on any of their graphs. The next day, the students could not agree on which variable should be labelled on the horizontal axis of each graph. The students made the following suggestions as to what each variable could be.

Student	Variable 1	Variable 2	Variable 3	Variable 4
Marcus	Substrate concentration	Temperature	pH	Enzyme concentration
Billy	Temperature	Substrate concentration	Enzyme concentration	pH
Voula	Enzyme concentration	Temperature	Substrate concentration	pH
Sheena	Temperature	pH	Enzyme concentration	Substrate concentration

Which student correctly identified all four variables on the horizontal axes?

A Marcus

B Billy

C Voula

D Sheena

Question 7

The active site of an enzyme may be denatured by heat. Excessive heat causes

A a change in the secondary and/or tertiary structure of the protein molecule.

B a change in the structure of the amino acid molecules in the protein.

C a change in the amino acid sequence in the peptide chains making up this region.

D particular amino acids to be substituted for less active amino acids.

Question 8

The following diagram represents an enzyme and its substrate molecules.

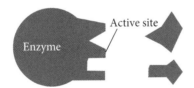

Which of the following diagrams represents a molecule that is likely to act as a competitive inhibitor of this enzyme molecule?

A

B

C

D

Short-answer questions

Question 9 (4 marks)

The appearance of a flower, its texture and scent, are an important part of the $4 billion international cut flower industry. Research has been undertaken to identify the key enzymes involved in each developmental stage to help identify plant varieties that have improved flower quality and post-harvest performance.

The cellular respiration occurring in flower cells is essential to keep the flowers looking fresh for long periods of time once the bloom has been harvested. The diagram below is a part of the pathway for cellular respiration in a flower bloom.

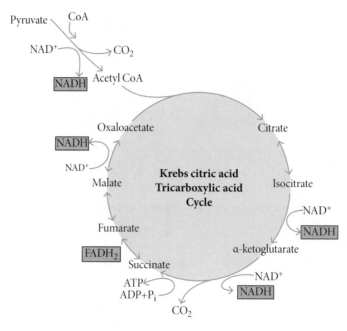

a   If a flower is cut, explain what will happen to this biochemical pathway if the reactants
    can no longer be supplied.   1 mark

b   i   If there is a mutation in the gene coding for succinate dehydrogenase (the enzyme
        responsible for catalysing the change from succinate into fumarate) resulting in a
        malformation of the enzymes active site, what would happen to this biochemical pathway?   1 mark

    ii Would this result in a long-lasting flower once it has been cut? Explain.   2 marks

Question 10 (2 marks)

Describe the role of ATP-synthase in oxidative phosphorylation.   2 marks

Question 11 (2 marks)

The enzyme citrate synthase is found in nearly all living cells and is involved in the first step of the Krebs
cycle. Citrate synthase is located in the mitochondrial matrix, but is encoded by nuclear DNA rather than
mitochondrial DNA, so is partly responsible for the regulation of this process. Since the activity of this
enzyme plays a significant regulatory role in the biochemical pathway, scientists wanted to investigate the
properties of this enzyme.

Describe what would happen to citrate synthase under the following conditions.

a   pH 2   1 mark

b   4°C   1 mark

Question 12 (5 marks)

Both alanine and ATP act as non-competitive inhibitors of pyruvate kinase, the enzyme that catalyses
the final step in the glycolytic pathway. The inhibition of pyruvate kinase allows cells to shut off the
breakdown of glucose when adequate amounts of end-product (ATP and alanine) are present.

a   i   Predict what would happen if the inhibition of pyruvate kinase did not occur.   1 mark

    ii What would happen if the inhibition of pyruvate kinase was not reversible?   1 mark

b Hexokinase functions in the first step of glycolysis to phosphorylate glucose, yielding glucose-6-phosphate. Glucose-6-phosphate is a potent non-competitive inhibitor of hexokinase, shutting off the process if plenty of glucose has already been broken down in the glycolytic pathway.

i Choose a diagram below that best represents this process.   1 mark

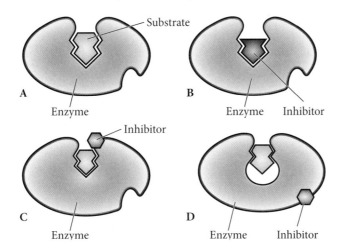

ii Discuss the difference between diagram B and diagram D.   2 marks

Question 13 (8 marks)

Amelia and Lara were completing an experiment to see what the effect of concentration of substrate and enzyme had on the rate of a reaction. They decided to use the enzyme amylase, because this enzyme catalyses the hydrolysis of starch into glucose ready for human cells to produce energy through cellular respiration. They set up the test tubes as described below.

Tube 1 – 10 mL starch solution (200 µg/mL) and 1 mL amylase solution (2.5 mg/mL)

Tube 2 – 10 mL starch solution (200 µg/mL) and 1 mL amylase solution (1.5 mg/mL)

Tube 3 – 10 mL starch solution (100 µg/mL) and 1 mL amylase solution (2.5 mg/mL)

Tube 4 – 10 mL starch solution (100 µg/mL) and 1 mL amylase solution (1.5 mg/mL)

Tube 5 – 10 mL starch solution (200 µg/mL) and 1 mL water

They incubated the test tubes at 37°C for 10 minutes and stopped the reaction by adding 1 mL of 1 M HCl solution to each tube.

a Using your knowledge about the effects of concentration on enzyme action, comparing the two tubes stated in parts i–iii, which of the tubes would have the least amount of starch when the reaction was stopped? Explain why in each case.

i Tube 1 or Tube 2   1 mark

ii Tube 1 or Tube 3   1 mark

iii Tube 4 or Tube 5   1 mark

b If Amelia and Lara wanted to investigate the rate of these reactions, how would the experiment need to be modified?   1 mark

c **i** Draw a graph comparing the predicted results in tubes 1 and 3.   2 marks

ii Draw a graph comparing the predicted results in tubes 3 and 4.   2 marks

Photosynthesis as an example of a biochemical pathway

Solutions start on page 224.

Multiple-choice questions

Question 1 ©VCAA | VCAA 2018 SA Q14 (adapted)

The diagram at the right shows the structures of a chloroplast, labelled R–W.

The light-independent reaction of photosynthesis occurs at

A T.   **C** V.

B U.   **D** W.

Question 2

Which of the following is a product of the light-independent reactions of photosynthesis?

A Carbon dioxide   **B** ATP   **C** Oxygen   **D** Glucose

Question 3 ©VCAA | VCAA 2018 SA Q15

Which one of the following diagrams correctly represents the inputs and outputs of photosynthesis?

A
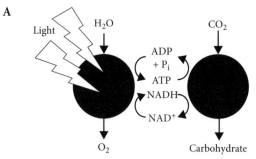

C
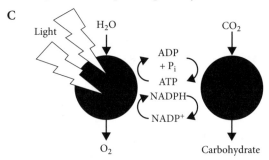

B
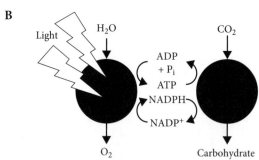

D
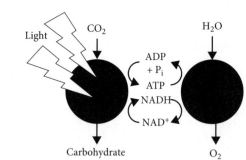

Question 4

NADPH is an important carrier of hydrogen used in

A glycolysis.

B the Krebs cycle.

C electron transport in mitochondria.

D the light-dependent stage in the chloroplast.

Question 5 ☑☑▢

In the grana of the chloroplast

A the splitting of water occurs during the light-dependent stage.

B the Calvin–Benson cycle occurs, releasing six carbon dioxide molecules.

C the electron transport chain occurs, using six carbon dioxide molecules to form glucose.

D glucose is created from a three-carbon sugar phosphate molecule for the storage of energy by the plant.

Question 6 ☑☑☑

Oxygen is a competitive inhibitor for Rubisco. In order to limit the exposure of Rubisco to oxygen, C_4 and CAM plants that live in hot, dry climates have evolved to

A open their stomata during the day to allow oxygen to escape readily.

B convert CO_2 into a four-carbon compound using a different enzyme.

C complete all stages of photosynthesis at night when the temperature is cooler.

D produce a compound to bind the oxygen so it does not interfere with carbon fixation.

Question 7 ☑☑▢

C_3 plants are named as such because

A Rubisco fixes carbon dioxide to a five-carbon ring, which then splits into two molecules containing three carbons each.

B they are only found in three places across the globe.

C they have three ways to fix carbon from carbon dioxide in the atmosphere, making them the most abundant types of plants on Earth.

D they produce three main products, whereas C_4 plants produce four.

Question 8 ☑▢▢

Photosynthesis is used by a plant to transform and store energy from the Sun. Which of the following variables can increase the rate of photosynthesis?

A Increased temperature

B Decreased concentration of carbon dioxide

C Increased concentration of oxygen

D Increased concentration of glucose

Question 9 ☑☑☑

The rate of photosynthesis increases when

A the length of the day becomes longer; therefore, more light is available for use in the Calvin–Benson cycle.

B a plant is moved from the direct sunlight into a shaded area.

C the level of atmospheric CO_2 increases because the rate of the light-independent reactions in the stroma increase.

D the stomata of the plant close and water is not lost from the leaves of the plant, resulting in an increased availability of water for photosynthesis.

Question 10 ⬤○○

Which of the plants below will have the fastest growth rate?

A *Betula pubescens* (downy birch) native to Iceland

B Desert grass trees native to outback Australia

C Lemon scented myrtle native to Queensland rainforests

D The saguaro cactus that grows 3 cm in the first 10 years of its life

Short-answer questions

Question 11 (4 marks) ©VCAA VCAA 2015 SA Q3 ⬤○○

Below is a diagram of a chloroplast.

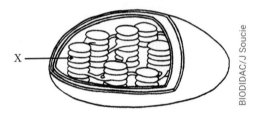

a Name the structure labelled X. 1 mark

b Complete the table by referring to the diagram above and your knowledge of photosynthesis. 3 marks

Name of the stage of photosynthesis that occurs at X		
Two input molecules that are required for reactions at X	1	2
Two output molecules that result from the reactions at X	1	2

Question 12 (5 marks) ©VCAA VCAA 2019 SB Q2 ⬤⬤○

a A chloroplast is surrounded by a double membrane.

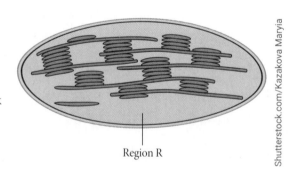

Region R

   i Name **two** molecules, as inputs for photosynthesis, that would need to diffuse from the cytosol of the plant cell across the chloroplast membranes and into the chloroplast. 1 mark

   ii Under high magnification, the internal structure of a chloroplast is visible. The diagram on the right shows part of this structure.

   A higher concentration of oxygen is found in Region R when a plant is photosynthesising compared to when it is not photosynthesising. Account for the differences in oxygen concentrations found in this region. 2 marks

b Describe the role played by each of the coenzymes NADPH and ATP in photosynthesis. 2 marks

Question 13 (2 marks) ●●○

In a series of experiments, leaf cells containing chloroplasts were exposed to light of varying wavelengths. The amount of oxygen produced in a two-hour period was measured and recorded. The results are shown in the table below.

	Sample 1 Wavelength 1	Sample 2 Wavelength 2	Sample 3 Wavelength 3
Amount of oxygen produced (mL)	130	100	15

a Which of the samples was exposed to green light?   1 mark

b Explain your answer to part **a**.   1 mark

Question 14 (4 marks) ●○○

a Pineapple, cacti and orchids are examples of CAM plants. In what type of environment are CAM plants found?   1 mark

b Maize and sugar cane are examples of C_4 plants. In what type of environment are C_4 plants found?   1 mark

c Name **two** adaptations that C_4 and CAM plants have developed to maximise the efficiency of photosynthesis.   2 marks

Question 15 (4 marks) ●●○

Since the 1860s, there have been nine major drought periods in Australian history. Describe the effect drought has on the rate of photosynthesis in Australian plants and how the amount of water affects the rate of photosynthesis in native plants and introduced species.   4 marks

Question 16 (5 marks) ●●●

In an investigation of photosynthesis, a laboratory class exposed leaves of a geranium plant to a source of radioactively labelled carbon as $^{14}CO_2$. The leaves were kept in water to prevent wilting. Prior to being placed in the beakers, some of the leaves were treated as shown in the table below. Light intensity and temperature were also varied during the experiment.

Leaf	Treatment	Light	Temperature
A	None	Dark	25°C
B	None	Sunlight	25°C
C1	Half leaf covered in foil	Sunlight	25°C
C2	Half leaf uncovered	Sunlight	25°C
D	None	Sunlight	70°C

These leaves were kept in the laboratory for 24 hours. The next day a number of discs were cut from each leaf, dried, and the amount of radioactively labelled product was measured.

The results of the experiment are shown in the table below. (Radioactivity is recorded in counts per minute of activity.)

Leaf	Counts per minute of activity
A	0
B	160
C1	10
C2	130
D	2

a Write a balanced equation for the process of photosynthesis.   1 mark

b **i** Give **two** explanations that could account for the presence of radioactively labelled product in leaf section C1.   2 marks

    ii How would you account for the low levels of product in leaf D?   2 marks

Cellular respiration as an example of a biochemical pathway

Solutions start on page 227.

Multiple-choice questions

Question 1 ◐●●

In what type of plant cells would you expect to find mitochondria?

A   All living cells

B   Cells that lack chloroplasts

C   None, because plant cells do not contain mitochondria

D   Root and stem cells only

Question 2 ©VCAA VCAA 2017 SA Q10 ●●●

The image at the right shows a three-dimensional diagram of an organelle found in eukaryotic cells.

The structure labelled Y is where

A   glucose enters glycolysis.

B NAD^+ is converted into NADH.

C   the majority of ATP is produced in the cell.

D   pyruvate is broken down, releasing carbon dioxide.

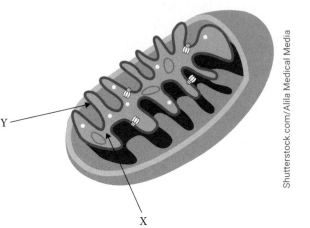

Y

X

Shutterstock.com/Alila Medical Media

Question 3 ©VCAA VCAA 2017 SA Q11 ●●●

An animal cell culture was exposed to radioactively labelled oxygen. The cells were then monitored for three minutes. After this time, the radioactively labelled oxygen atoms would be present in which cellular chemical?

A   Adenosine triphosphate

B   Carbon dioxide

C   Glucose

D   Water

Question 4 ©VCAA VCAA 2018 SA Q9 ●●●

Which of the following gives the inputs and outputs of the electron transport chain in an animal cell?

	Inputs	Outputs
A	NADH, ADP, oxygen, P_i	ATP, NAD^+, water
B	NADH, ADP, water, P_i	ATP, NAD^+, oxygen
C	NAD^+, ADP, oxygen, P_i	NADH, ATP, water
D	NADPH, ADP, water, P_i	$NADP^+$, ATP, oxygen

Question 5 ◐●●

A product of anaerobic fermentation in yeast cells is

A   glucose.

B   lactic acid.

C   ethanol.

D   water.

U3 – AOS2 – EXAM PRACTICE

Question 6

Which of the following processes results in a net gain of ATP molecules?

A Anaerobic respiration

B Protein production

C Active transport

D Muscle contraction

Question 7

In times of strenuous exercise, more energy is required by the body than when it is at rest. Which of the following does NOT increase the rate of aerobic cellular respiration in human cells?

A Heavy breathing, which increases the concentration of oxygen in the cytosol of cells

B The slight increase in the temperature of the body's cells as the body heats up

C Heavy breathing, which decreases the concentration of carbon dioxide in the cytosol of cells

D A decrease in the concentration of glucose in the cytosol of the cells

Use the following information to answer Questions 8 and 9.

The graph below shows the net output of oxygen in spinach leaves as light intensity is increased. Temperature is kept constant during the experiment.

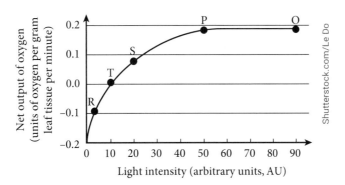

Question 8

At what point on the graph is the rate of cellular respiration equal to the rate of photosynthesis?

A R

B T

C S

D Between points P and O

Question 9 ©VCAA VCAA 2017 SA Q14 (adapted)

Which one of the following conclusions can be made based on the graph?

A Between points P and O, photosynthesis has stopped and no more oxygen is being produced.

B Between points P and O, enzyme concentrations in the cells is preventing the rate of photosynthesis from increasing further.

C The optimum light intensity is at point T.

D The optimum light intensity is at 70 AU.

Short-answer questions

Question 10 (5 marks)

A biology student was asked to draw a flow chart that summarised his understanding of cellular respiration. The chart over the page is what he produced.

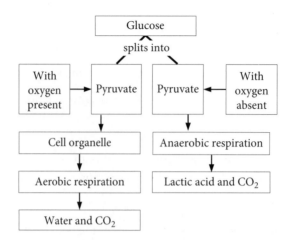

This diagram was seen as inadequate by his teacher, who made a number of criticisms and pointed out areas where important detail was left out.

a What is the name of the process in which the two pyruvate molecules are formed?   1 mark

b The student has made an error in his products of anaerobic respiration.
What should he have included as products of anaerobic respiration?   1 mark

c The student did not name the organelle in which cellular respiration occurs.
What is the name of this organelle?   1 mark

d The student's diagram makes no mention of the release of energy in these reactions.
Add to his diagram to show where, and how much, ATP is produced for use by the cell.   2 marks

Question 11 (6 marks)

Copy and complete the table below.

Biochemical process	Location	Inputs	Outputs	No. of ATP produced
Glycolysis				
Krebs cycle				
Electron transport chain				

Question 12 (7 marks) ©VCAA VCAA 2013 SB Q1

Yeast is a single-celled, microscopic fungus that uses sucrose as a food source. An experiment was carried out to investigate cellular respiration by a particular species of yeast. Yeast cells were placed in a container and a sucrose solution was added. An airtight lid was placed on the container. The percentages of oxygen and ethanol in the container were recorded over a one-hour period. The experiment was carried out at room temperature. The results are shown in the following table.

	Percentage of oxygen	Percentage of ethanol
At the start of the experiment	21	0
At the end of the experiment	18	4

a Explain any changes that have been observed in oxygen and ethanol levels within the airtight container.   2 marks

b Levels of carbon dioxide were also monitored during the experiment. Predict whether the carbon dioxide concentration inside the airtight container would increase, stay the same or decrease within the time the experiment was carried out. Explain the reasoning behind your prediction.   2 marks

c   Scientists are looking at ways to increase the efficiency of photosynthesis in plants, including the way in which carbon dioxide is captured.

   i   Name the stage of photosynthesis in which carbon dioxide is captured.   1 mark

   ii The stage of photosynthesis in which carbon dioxide is captured requires other inputs. Name **two** other inputs and describe the role played by each in this stage of photosynthesis.   2 marks

Question 13 (9 marks)

A biologist carried out a series of experiments to investigate the rate at which phosphate ions were taken up by yeast cells. Yeasts are capable of aerobic respiration, but when deprived of oxygen they can produce energy by fermentation. Three test tubes were set up, each containing the same amount of yeast and the same concentration of phosphate ions. The tubes varied in the presence of glucose and/or oxygen.

The test tubes were kept in the dark, and the amount of phosphate taken up by the yeast was measured over a three-hour period. The results are shown in the table below.

	Contents of each tube and phosphate uptake		
Time from start of experiment (hours)	TUBE 1 Glucose absent Oxygen present	TUBE 2 Glucose present Oxygen absent	TUBE 3 Glucose present Oxygen present
0	0	0	0
1	0	15	30
2	0	25	55
3	0	25	80

a   Under what conditions was the yeast able to take in the most phosphate?   1 mark

b   i   By what process was the phosphate taken up by the yeast?   1 mark

   ii Explain your answer to part **b i**.   1 mark

c   i   What metabolic process must have been occurring in tube 2 throughout the experiment to allow the uptake of phosphate?   1 mark

   ii Write an equation for this process.   1 mark

d   i   What metabolic process was occurring in tube 3?   1 mark

   ii Write an equation for this process.   1 mark

e   Throughout the experiment the rate of uptake of phosphate remained higher in tube 3 than in tube 2. How would you account for this?   2 marks

Question 14 (4 marks)

a   Where do anaerobic processes occur in a cell?   1 mark

b   How many ATP are produced during anaerobic fermentation?   1 mark

c   Which is the more efficient process, aerobic or anaerobic respiration? Explain your answer.   2 marks

Question 15 (8 marks)

Dion made a New Year's resolution to get fit. As he had not been running for several months, he decided he would go for a run every morning to achieve his goal. During his first run of the year, he found that within the first 2 minutes he was breathing rapidly and sweating profusely, and his body temperature was much higher than usual. The next day his legs were very sore.

a Why was Dion breathing rapidly within the first few minutes of his run? Comment on the rate of cellular respiration. 2 marks

b Why did Dion's temperature increase and how does this affect aerobic cellular respiration? 2 marks

c What would happen if Dion's temperature was raised too much? 1 mark

d Why were Dion's legs sore the day after his run? 2 marks

e Write a word equation for the reaction discussed in part **d**. 1 mark

Biotechnological applications of biochemical pathways

Solutions start on page 229.

Multiple-choice questions

Question 1

Which of the statements below is a valid reason for the use of CRISPR-Cas9 technology in plants?

A The need to increase crop productivity due to human population increases and changes in global climates

B The need to grow tomatoes that are blue in colour

C The need to decrease the growth rate of cultivars with less desirable qualities

D The need to increase the length of time a crop takes to mature

Question 2

Which of the following is NOT a way to improve crops using CRISPR-Cas9 technology?

A Resistance to biotic stresses such as viruses, bacteria, fungi and insects

B Resistance to abiotic stresses such as drought, flooding, heat and cold

C Improvement of fruit quality such as size, texture and nutrient content

D Improvement of soil composition such as the retention of nutrients and water

Question 3

An increase in population is resulting in a need for more agricultural land to grow crops. However, there is also an increase in the global need to maintain and increase natural forests as atmospheric carbon dioxide levels rise. Which of the following is NOT a solution to this problem?

A Modify light harvesting pigments so that the plant increases biomass, resulting in plants that grow larger in smaller land areas

B Modify rice to use the C_4 plant pathway so that the crops can be grown in hot and dry environments

C Introduce carbon dioxide chambers into photosynthetic cells to allow for higher carbon dioxide concentration within cells, leading to higher efficiency of biomass production

D Modify animals used in agriculture to increase muscle mass so that they feed more people

Question 4

Biofuels are

A fuels derived from the breakdown of fibrous plant waste.

B the use of animals to generate electricity.

C any fuel used by farmers in the production of large-scale crops.

D the biomolecules used to produce energy within living organisms.

Question 5 ▢▢▢

Which of the following does NOT involve anaerobic fermentation of biomass for the production of biofuels?

A The fermentation of glucose by yeast to produce ethanol and carbon dioxide

B Collecting the manure from animals for bacterial digestion, producing methane

C Increasing oxygen supply for bacteria to convert biomass into a usable fuel source

D The bacterial digestion of waste plant matter to produce bioethanol

Question 6 ▢▢▢

Microalgae are great candidates for sustainable production of biofuels. This is because they

A are small and grow quite slowly, taking carbon dioxide out of the environment.

B can have a high biomass that can be converted into or used directly as fuel.

C don't photosynthesise, so reactants for their growth do not have to come from the atmosphere.

D have only one strain, so are easily produced.

Short-answer questions

Question 7 (4 marks) ▢▢▢

a Why is the use of CRISPR-Cas9 considered a better way to improve photosynthetic efficiencies than transgenesis?   2 marks

b How does this technology compare with the natural environmental adaptation to different environments? Why is it important now in the 21st century?   2 marks

Question 8 (10 marks) ▢▢▢

a Name the two types of fuels that can be generated by anaerobic fermentation.   2 marks

b Write the word and chemical equation for the generation of bioethanol.   2 marks

c Why are biofuels considered sustainable and carbon neutral?   2 marks

d Draw the cycle of carbon as it is taken in by plants and then released back into the atmosphere when used as a biofuel. Include an example of a use for the biofuel in your response.   4 marks

Question 9 (9 marks) ▢▢▢

In China, many large farms in rural areas use crop and animal waste in fermenters to generate biofuels for energy use on the farm.

a Name **two** uses for the biofuels generated.   2 marks

b Give **three** advantages of generating biofuels from animal and plant waste.   3 marks

c Some farmland can be used specifically to produce crops for the production of biofuels. Discuss the implications of growing crops specifically for this purpose and not just using waste products.   4 marks

Question 10 (7 marks) ▢▢▢

a **i** List **four** raw reactants used to produce biogas.   2 marks

   ii List some uses for biogas once it has been generated from biomass, as well as uses for the organic matter left over in the anaerobic digester.   2 marks

b What are the requirements within the biomass digester in order to produce biofuels?   3 marks

UNIT 4

HOW DOES LIFE CHANGE AND RESPOND TO CHALLENGES?

Chapter 3 Area of Study 1
How do organisms respond to pathogens?

Area of Study summary

This unit explores some of the challenges faced by organisms to survive and reproduce. These challenges include environmental pressures and disease, which can lead to evolution.

Unit 4, Area of Study 1 explores the biology behind the immune response of humans, and how, at a molecular level, it is able to adapt to protect itself from invading pathogens. The challenges and strategies that are used to treat and prevent outbreaks of disease are explored, including how vaccinations provide immunity and how technological advances can assist with managing and treating diseases. The challenges and strategies in the prevention of disease in a globally connected world are also covered.

Acquired biological knowledge from this area of study can be applied to bioethical issues related to disease and establish how life changes and responds to challenges, such as allergic reactions, development of immunotherapy strategies, patterns and evidence of evolutionary relationships and conservation planning.

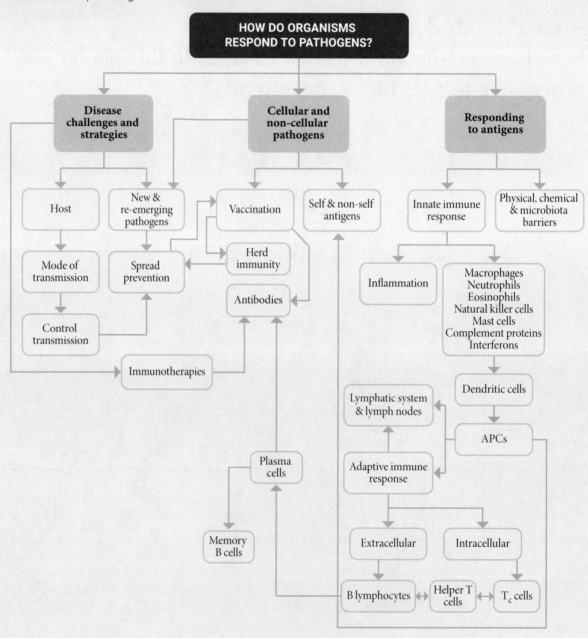

Area of Study 1 Outcome 1

On completing this outcome, you should be able to:

- **analyse** the immune response to antigens
- **analyse** the different ways that immunity can be acquired
- **evaluate** the strategies and challenges in the treatment of disease.

The key science skills demonstrated in this outcome are:

- **analyse**, **evaluate** and **communicate** scientific ideas
- **create** evidence-based arguments and draw conclusions.

© (2021). The Victorian Curriculum and Assessment Authority (VCAA). Used with permission.

3.1   Responding to antigens

Antigens are molecules on the self-surface membrane that act as markers to the body. The body can recognise antigens as **self antigens** or **non-self antigens**. Once non-self-antigens invade the body, a set of responses is elicited in order to protect against infection and disease.

3.1.1 The first line of defence: barriers to pathogenic infection in plants and animals

Infection can occur when viruses, bacteria and other microbes gain entry to a host and begin to multiply. When the cells in the organism are damaged, this leads to illness and, in some cases, death of the host. **Pathogens** can gain entry through skin contact, bodily fluids, airborne particles or through contact with contaminated food or water. Plants and animals have defences to prevent invasion of pathogens that include physical, chemical and microbiotic barriers. These are collectively known as the **first line of defence** and they are summarised below.

Barrier	Preventative mechanisms in animals	Preventative mechanisms in plants
Physical (to prevent entry of pathogens)	Intact skin acts as a barrier to pathogens, preventing entry into the body. The skin also produces biochemical defences that prevent growth of pathogens. Cilia, eyelashes and hairs help to trap and block pathogens from entering the body through openings.	Hairs, spines and prickles on the surface of leaves or stems may stop small insect vectors from coming into contact with a plant's tissue. A waxy cuticle provides a barrier to the entry of pathogens. Thick barks and closing of stomata also prevent entry to pathogens. In some species, infection with certain pathogens leads to the formation of galls, impenetrable capsules in which the pathogen is trapped.
Chemical (to inhibit growth or development of pathogens)	Mucous membranes produce mucus to trap and protect the internal structures such as the trachea, vagina and urethra. Tears, saliva, and mucus wash away pathogens. They also contain lysozyme, which lyse bacterial cell walls. Acid in the stomach creates a hostile environment that destroys pathogens and prevents their growth.	Phenolic compounds in plant tissue are oxidised after the tissue has been damaged, for example, by an insect bite. This produces toxic chemicals that provide a barrier to the entry of fungi or bacteria. Plants produce antibiotic chemicals (known as phytoalexins) in response to tissue damage. These can kill invading microbes. The cells of some plants can produce enzymes such as chitinases, which digest the cell walls of fungi. Cell walls of fungi are made from chitin.

››

Barrier	Preventative mechanisms in animals	Preventative mechanisms in plants
Chemical (continued)		Some plants produce insect hormones. When insect larvae eat part of these plants, their life cycle is disrupted.\n\nSome plants produce toxins that deter browsing animals. This contributes to defence against pathogens because fewer bites means fewer wounds for the entry of parasites.
Microbiota (to prevent the growth or colonisation of pathogenic micro-organisms)	The normal flora that colonise humans help to prevent pathogens by competing for attachment sites on the skin, gastrointestinal tract and vagina, and obtaining essential nutrients. They can also produce substances that kill off or inhibit the growth of other bacteria. Normal flora can also stimulate the production of natural antibodies.	Primarily soil microbes are believed to produce antimicrobial peptides.

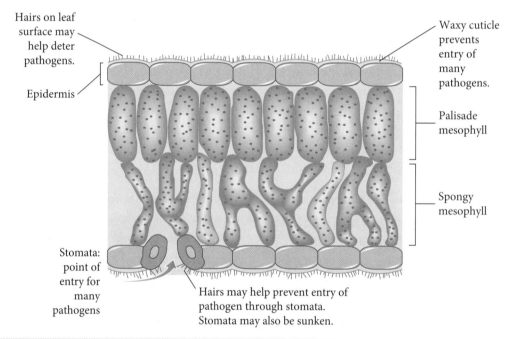

Hairs on leaf surface may help deter pathogens.

Waxy cuticle prevents entry of many pathogens.

Epidermis

Palisade mesophyll

Spongy mesophyll

Stomata: point of entry for many pathogens

Hairs may help prevent entry of pathogen through stomata. Stomata may also be sunken.

A cross-section of a typical leaf, highlighting some of the physical plant barriers to pathogens

Hint

If a question asks you to recall a cell from the innate immune response and explain its role, mentioning molecules will be an incorrect response!

3.1.2 The second line of defence: the innate immune response

The **innate immune response**, sometimes referred to as the second line of defence, is non-specific, which means it cannot differentiate between different types of pathogens. The defence mechanisms are triggered once a pathogen gets past the first line of defence. The innate response occurs within hours and it will always produce the same set of responses to try to protect the body.

The second line of defence can be broken down into cellular defences and molecular defences. The cellular defences include cells such as **mast cells** and neutrophils. Molecular defences include proteins that contribute to the body's innate defences against pathogens.

Cellular defences

Leukocytes

Leukocytes are white blood cells that help to clear the body of infection. They can be categorised into granulocytes and monocytes. Granulocytes contain enzymes in granules that can damage or digest pathogens. Monocytes differentiate into either macrophages or dendritic cells and these are phagocytotic.

Leukocyte	Characteristics	Role	
Macrophage	Monocyte	Phagocytic cell that digests foreign pathogens Stimulates responses of other immune cells	 Alamy Stock Photo/Cultura Creative RF
Dendritic cell	Monocyte	Antigen-presenting cell to trigger adaptive immunity	 Alamy Stock Photo/agefotostock
Neutrophil	Granulocyte	Phagocytic cell that makes up approximately 50–60% of all leukocytes Releases toxins to kill or inhibit bacteria and fungi	 Science Photo Library/David M. Phillips
Eosinophil	Granulocyte	Releases toxins that kill bacteria and parasites Can cause tissue damage	 Alamy Stock Photo/Science Photo Library
Mast cell	Granulocyte	Stimulates vasodilation and induces inflammation through releasing histamines Recruits macrophages and neutrophils	 Alamy Stock Photo/Science Photo Library

Inflammation

Inflammation is a reaction caused by the release of **histamines** from mast cells. The release is triggered when tissue becomes damaged. Histamine increases the permeability of capillaries by causing vasodilation. This allows immune cells to access infected tissue through the small gaps between the cells that make up the capillary vessel wall. Inflammation leads to redness, swelling, heat and pain.

Phagocytosis

Phagocytosis is the engulfment and digestion of pathogens through endocytosis by specialised cells such as macrophages. The term phagocytosis translates to 'eating cell'. Macrophages in the blood concentrate at the sites of infection due to the release of chemicals, such as histamine. The steps involved in phagocytosis are outlined below.

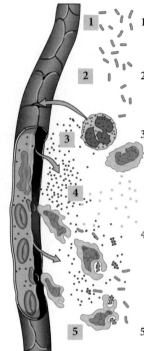

1 Bacteria enters tissue following injury or infection.

2 Histamine is released by mast cells and macrophages release cytokines.

3 Histamine triggers the blood vessels to widen. Gaps form within the capillary vessel and it becomes permeable, allowing tissue fluid and proteins to enter surrounding tissue.

4 Cytokines recruit phagocytes to the area. They travel in the blood and gain access to the surrounding tissue through the leaky capillary vessel.

5 Phagocytes engulf bacteria.

The process of inflammation allows phagocytes entry to the site of damage.

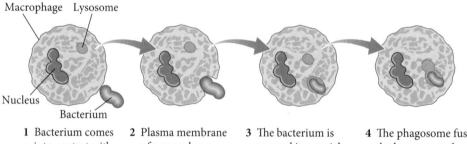

1 Bacterium comes into contact with macrophage.

2 Plasma membrane of macrophage envelopes the bacterium and engulfs it.

3 The bacterium is encased in a vesicle called the phagosome.

4 The phagosome fuses with the lysosome to become the phagolysosome, which now contains digestive enzymes and low pH that destroy the bacterium.

The steps involved in phagocytosis

Fever

Fever is an inflammatory response that causes a rise in body temperature. Pathogens, which have become ingested through phagocytosis, release toxins known as pyrogens. These **pyrogens** stimulate the release of **prostaglandins** in the brain, which triggers the body's thermoregulating control centre, the hypothalamus, to coordinate the body to generate and retain heat through positive feedback. The higher temperature will reduce the growth rates of pathogens and stimulate leukocytes to destroy them.

9780170479431

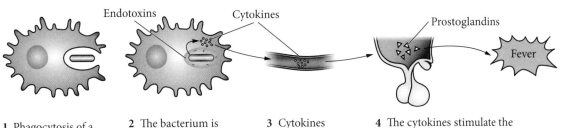

1 Phagocytosis of a bacterium by a macrophage

2 The bacterium is digested in a vacuole, releasing endotoxins. The endotoxin triggers the macrophage to produce cytokines.

3 Cytokines released by the macrophage travel in the blood to the hypothalamus in the brain.

4 The cytokines stimulate the hypothalamus to produce prostaglandins, which raise the body's temperature, resulting in a fever.

The steps involved in fever

Molecular defences

Complement proteins

Complement proteins are named as such because as they are complementary to the antibody response of the adaptive immune system. Complement proteins include approximately 20 types of soluble proteins that travel through the blood and become activated by pathogens, triggering a cascade where the pathogen becomes coated in complement proteins. This serves a number of functions, including the recruitment of **phagocytes** through **opsonisation** of pathogens, where the complement proteins tag foreign invaders for destruction. They are also able to increase inflammation and some proteins are able to destroy cell membranes by creating open pores, causing their contents to leak.

Cytokines

Cytokines are proteins that act as signalling molecules in an immune response. They coordinate inflammation and immune responses. **Interleukins** and **interferons** are types of cytokines.

Interleukins act as messenger molecules between immune cells. Interferons are released by infected cells, triggering nearby cells to reduce their susceptibility to pathogens. Interferons recruit natural killer (NK) cells, which target and destroy infected cells.

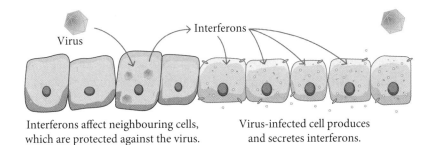

Interferons affect neighbouring cells, which are protected against the virus.

Virus-infected cell produces and secretes interferons.

Interferons act on healthy neighbouring cells to protect them from viral entry and prevent the virus from replicating.

Natural killer (NK) cells

Natural killer (NK) cells are granular lymphocytes able to insert their granules that contain cytotoxic chemicals into infected cells. Perforins and enzymes create pores in the membrane of the infected cell and induce apoptosis, killing the cell. NK cells can also produce cytokines, recruiting phagocytes to clean up. NK cells patrol the body and are always in contact with cells. Whether or not they kill depends on receptors becoming activated by receptors on the body's cells.

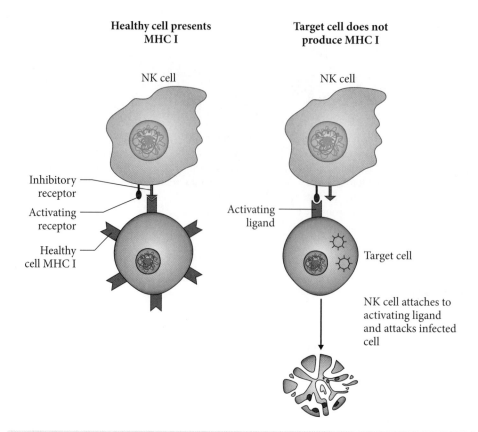

NK cells recognise infected cells and induce apoptosis within the cell.

3.1.3 The initiation of an immune response

All cells contain genetic material. Some of the genetic material will be expressed during translation as specific proteins known as **major histocompatibility complexes (MHC)**. These are embedded on the cell surface membrane of all nucleated cells in the body (so *not* red blood cells).

As these proteins are translated from genes, the variations of alleles between people result in each individual having unique MHCs. The role of MHCs is to display **antigens**: protein molecules that can be identified by the body's immune system. The body's immune cells can bind to MHC class I (MHC I) in order to check whether antigens are self or non-self. Non-self antigens are found on the surface of foreign invaders such as viruses and bacteria. Identifying invaders through the detection of non-self antigens stimulates an immune response.

NK cells can be activated to kill when there is no MHC I receptor on a cell. As cancerous cells and infected cells often lose their MHC I receptors, NK cells can be primed to destroy them.

The unique differences between MHC I markers in individuals result in a range of peptides being able to bind to them. This can lead to the variances in infection rates and autoimmune diseases within a population. It also has implications for tissue and organ donors, because it is almost impossible to have a perfect match with another individual. In fact, this where MHCs got their name: as *Histo* is derived from the Greek word for 'tissue', and 'compatibility' means to 'exist in harmony'.

MHC class II (MHC II) markers are found on some cells of the innate immune system, including some we have looked at from the innate immune response: macrophages and dendritic cells. These phagocytotic cells digest foreign invaders and bind some of the digested molecules to bind to MHC II markers, enabling presentation of these non-self antigens to cells such as T lymphocytes for recognition and destruction.

9780170479431

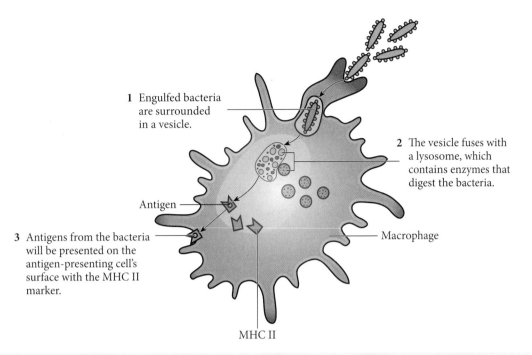

1 Engulfed bacteria are surrounded in a vesicle.

2 The vesicle fuses with a lysosome, which contains enzymes that digest the bacteria.

Antigen

3 Antigens from the bacteria will be presented on the antigen-presenting cell's surface with the MHC II marker.

Macrophage

MHC II

The steps involved in presenting MHC II antigens

Allergens

Allergens are not to be confused with antigens. Allergens are naturally occurring proteins that are usually harmless; however, if they are recognised by the body as non-self, then they will trigger an inflammatory response. In contrast, an antigen is harmful and will cause an adaptive immune response. An allergen can be any protein that the body responds to. Common allergens include pollen, dust, nuts and shellfish.

In people who suffer allergic responses, initial exposure to an allergen will result in the production of IgE antibodies. The production is triggered when an allergen's antigen encounters a B cell in the lymph or in the secondary lymphoid tissue. The B cell will then differentiate and proliferate into plasma cells, which will produce IgE antibodies specific to the allergen. The IgE antibodies then bind to mast cells at the constant region, embedding themselves into the plasma membrane. This primes the mast cells to become sensitised to the allergen. An allergic response will *not* occur at this stage.

On secondary exposure to the allergen, the allergen will bind to the specific IgE antibodies on the mast cell. For the mast cell to activate and release histamine, two or more IgE molecules need to be cross linked. The cross-linking results in a mass release of histamine from the mast cell. The histamine will result in inflammation, which is why allergic reactions involve redness, heat, swelling, watery eyes, sneezing and pain.

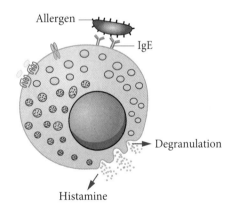

Allergen

IgE

Degranulation

Histamine

The cross-linking of IgE antibodies on mast cells caused by an allergen, triggering histamine release by a mast cell

Cellular and non-cellular pathogens

The table on the next page outlines the two types of pathogens: cellular (or living) and non-cellular (non-living). These pathogens, except for prions, would be identified by the body as non-self and would trigger an immune response.

Pathogen	Structure	Examples of diseases
Prions	Protein molecule Non-cellular	Creutzfeldt–Jakob disease Kuru Bovine spongiform encephalopathy (mad cow disease)
Viroids	A short strand of RNA with no protein coat Non-cellular Viroids occur in plants	Potato spindle tuber viroid Coconut cadang-cadang viroid Apple scar skin viroid
Viruses	Nucleic acid surrounded by a protein coat Some are enveloped in a modified membrane Non-cellular	Human Immunodeficiency Virus (HIV) SARS-CoV-2 (COVID-19) Influenza
Bacteria	Prokaryotic cells: lack membrane-bound organelles Contain a single round chromosome Reproduce by budding or binary fission	Salmonella Typhoid Tuberculosis Pertussis (whooping cough) Syphilis
Fungi	Unicellular or multicellular organisms Consist of eukaryotic cells with cell walls composed of chitin Reproduce by budding, fragmentation and spores	Athlete's foot Ringworm Thrush
Protozoa	Unicellular, colonial or simple multicellular organisms that consist of eukaryotic cell(s) Are animal-like protists: ingest food Reproduce by budding or binary fission	Malaria Toxoplasmosis Cryptosporidiosis
Worms	Multicellular, eukaryotic, specialised for a parasitic way of life Mouthparts may be modified to form hooks; digestive systems are simple, and they reproduce sexually to produce numerous offspring	Ringworm Hookworm Tapeworm

3.2 Acquiring immunity

Have you ever heard of a chicken-pox party? Instead of walking away with a lolly bag, you go home with a case of the pox. The rationale is that exposing young children to an individual infected with chicken pox allows their body to develop immunity after infection.

Immunity is the ability to resist harmful organisms from causing infection or disease. There are multiple types and ways in which a person can develop immunity, including naturally, through coming into contact with a pathogen, or artificially, through a vaccine.

3.2.1 The role of the lymphatic system in the immune response

The lymphatic system is a network of vessels, organs and lymph nodes that help to transport white blood cells in the lymph system throughout the body. It is responsible

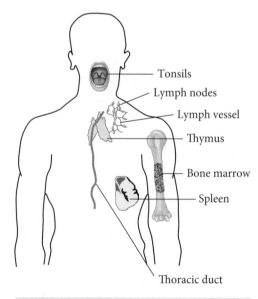

The vessels, organs and nodes that make up the lymphatic system

for the development of lymphocytes and connects the lymph nodes to the bloodstream. Lymph is formed from fluid that is forced out of the gaps in capillaries. Red blood cells are too large to fit through the gaps, but the fluid and nutrients present in blood will move out of the blood into the surrounding tissue, to bathe cells with nutrients. This tissue fluid will then either re-enter the bloodstream or enter the lymphatic system. Any tissue fluid that enters the lymphatic system becomes known as lymph.

The bone marrow is responsible for producing specialised leukocytes known as **lymphocytes**. There are two types of lymphocytes in the adaptive immune response: B lymphocytes (B cells) and T lymphocytes (T cells). B cells are named in reference to bone marrow, which is where they develop and mature. T cells are also produced in the bone marrow, but they travel to the thymus to mature, which is where the T comes from.

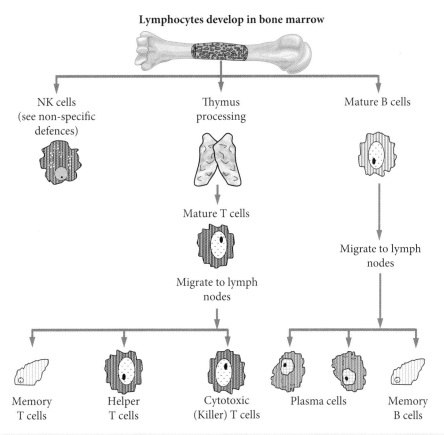

The life cycle of lymphocytes, including the site of production and maturation for B and T lymphocytes

Lymph carries any foreign substances, such as pathogens, cancer cells and damaged cells, to the lymph nodes for disposal. All substances transported through the lymph will pass through one or more lymph nodes, so antigen-presenting cells (APCs) such as macrophages, dendritic cells and B cells travel to the lymph node. These APCs have MHC II markers to present non-self antigens. The gathering of leukocytes at lymph nodes explains why lymph nodes become swollen and painful when you are experiencing a bad infection.

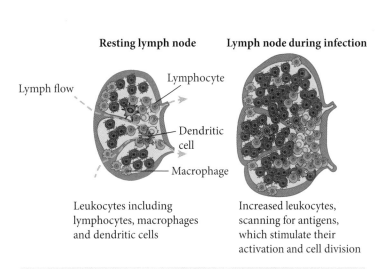

Leukocytes including lymphocytes, macrophages and dendritic cells

Increased leukocytes, scanning for antigens, which stimulate their activation and cell division

During an infection, the lymph nodes will swell due to increased activity.

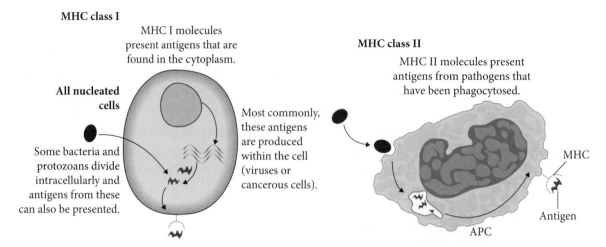

APC cells presenting foreign antigens to helper T cells

3.2.2 The third line of defence: the adaptive immune response

The adaptive immune response involves B and T lymphocytes. It is also known as the body's third line of defence, or 'specific immunity', because the immune cells coordinate a defence against specific antigens by producing antibodies. The adaptive immune response can be split into two categories: the humoral immune response and the cell-mediated response.

Antibody structure

Antibodies, also known as **immunoglobulins**, are quaternary proteins because they are made up of four polypeptide chains arranged in a 'Y' shape: two heavy chains and two light chains. The heavy chains are on the inside of the molecule, and the light chains are on the outside, with three sites holding the chains together using disulfide bonds between the cysteine amino acids. At the ends of the arms, there are regions that will vary between antibodies. Each antibody has two binding sites that are the complementary shape for a specific antigen.

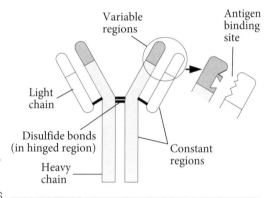

The general structure of an antibody

Types of antibodies

There are five types of antibodies or immunoglobulins. They are produced by B plasma cells. IgG are the most common and are able to cross the placenta. IgA are found in secretions such as tears, and IgE bind to mast cells and trigger the release of histamines in response to allergens.

The humoral immune response and the role of antibodies

The **humoral immune response** is facilitated by B cells and uses antibodies to fight extracellular

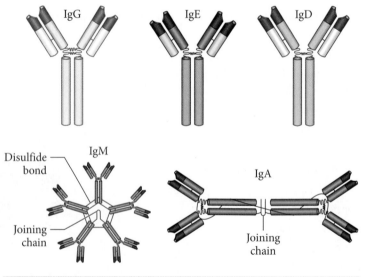

Antibody structure and the different classes

non-self antigens (pathogenic antigens found outside of the cell), such as those found on bacteria. As mentioned previously, B cells are produced in the bone marrow but circulate through the body in the blood, tissue fluid and lymph, in which they may encounter an antigen they recognise. Random genetic rearrangements or mutations during initial production allow for a diverse range of lymphocyte receptors to be generated. Recognition of a complementary antigen by a B lymphocyte (by direct binding) or T lymphocyte (via MHC) (**clonal selection**), triggers an impressive response in the selected cell causing it to divide rapidly, forming many copies or clones of itself and the specific antigen receptor it carries (**clonal expansion**). These clones can be one of two types of cells: effector cells (antibody-secreting

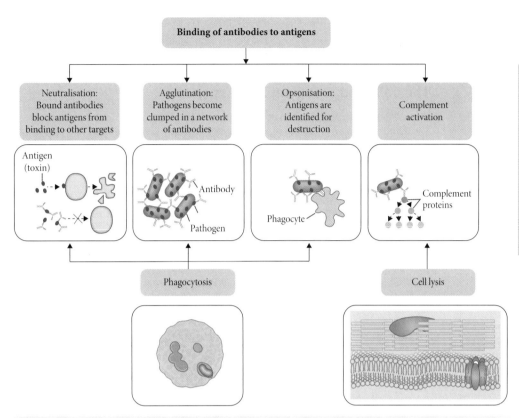

Protective mechanisms and consequences of antibody–antigen binding

B cells, or helper and cytotoxic T cells) or memory cells that confer long-term immunity. Alternatively, an antigen-presenting cell can activate a helper T cell, which in turn activates the corresponding B cell to undergo clonal expansion.

The army of clones are known as **plasma B cells** and they release specific antibodies into the bloodstream. These antibodies will be complementary to the antigen that was presented by the B cell to the helper T cell.

When the antibodies bind to the antigen it will result in either **neutralisation**, which prevents any further interaction with the body's own cells; opsonisation, where the antibody acts as a marker alerting macrophages to destroy the cells through phagocytosis; or complement, where the antibodies activate the complement system to destroy the marked cells through lysis. **Agglutination** results in an antibody–antigen complex that makes it easier for phagocytes to clear away pathogens.

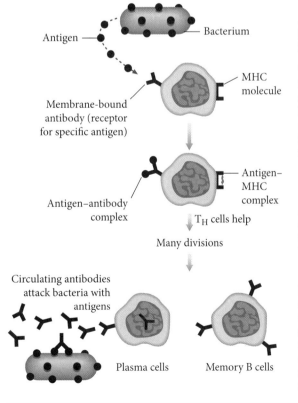

An immature B cell carries its specific antibody mounted on its membrane.

The immature B lymphocyte encounters the specific antigen recognised by its antibody. The antigen and antibody bond and form an antigen–antibody complex. This is assisted by secretions from T_H cells after they have recognised the antigen–MHC complex.

A series of reactions occur and the activated B cell divides many times. This is assisted by cytokines from T_H cells.

Some of the activated B cell clones become plasma cells that produce and secrete large numbers of antibody molecules that help to destroy the pathogen. Other activated B cell clones become memory cells to be ready for future encounters with the same antigen.

The process of B cell activation by a helper T cell, resulting in clonal expansion

Memory B cells are also created through clonal expansion. Memory B cells provide future immunity because they can survive for years and, as they have a receptor for a specific antigen, will trigger a rapid antibody-mediated immune response if the body becomes re-infected.

The cell-mediated immune response

The **cell-mediated immune response** becomes activated against non-self antigens that are intracellular (found within the cell), as antibodies are not effective against viruses that have entered the body's cells, or against cancer cells or transplanted tissue.

T lymphocytes, or T cells, regulate the cell-mediated response. T cells do not recognise free antigens, so they must bind to the body's cells. There are two types of T lymphocytes you need to be familiar with:

- Helper T cells (or T_H cells): These are regulatory cells and have cell-differentiated glycoprotein receptors known as CD4. CD4 receptors allow T_H cells to bind to MHC II markers.

- **Cytotoxic T cells** (or T_C cells): These cells are killer cells and have CD8 receptors that allow T_C cells to bind to MHC II markers.

Helper T cells become activated when they bind to MHC II markers presenting non-self antigens. The T_H cell secretes cytokines that stimulate B cells to undergo clonal expansion, producing antibodies for humoral immunity. The helper T cells are also stimulated to increase in number and help recognise the foreign antigen.

Cytotoxic T cells can bind to infected cells on the MHC I marker. Cytokines released by helper T cells activate T_C cells to proliferate and produce many more T_C cells that recognise the non-self antigen. T_C cells undergo **degranulation**, releasing **perforin**, a chemical that lyses the plasma membrane and kills the cell. Or, they can initiate cell-mediated **apoptosis** of the infected cell by binding to **death receptors**. This is similar to NK cells; however, T_C cells require activation and are able to recognise antigens, thus are included in the adaptive immune response.

Once an infection has cleared, the immune cells will die off. However, as with B cells, memory T cells are created in order to store information of antigens from previous infections. The **memory cells** can rapidly proliferate to vast numbers if they are exposed to an antigen they have seen before, which provides immunity to past infections.

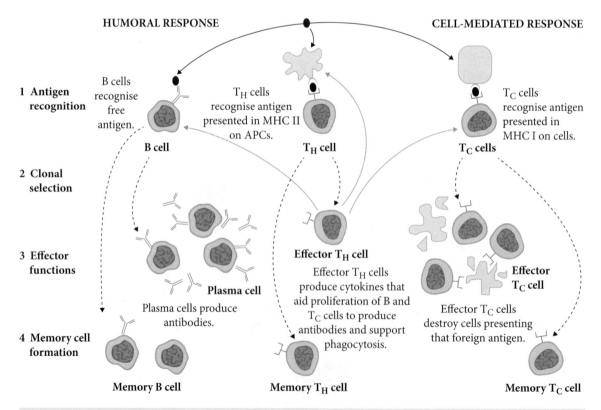

Comparison of the humoral and cell-mediated immune responses

3.2.3 Types of immunity

To be immune is to have an adequate biological defence to avoid an infection or disease caused by a pathogen. Humans are born with innate, or non-specific, immunity and, as we become exposed to more pathogens, we acquire active immunity through the production of antibodies.

Active immunity occurs when the body is stimulated to produce antibodies. This can be natural, through infection of a pathogen, or it can be induced artificially by exposing a person to a weakened or attenuated version of an antigen through **vaccination**. In both instances, the body is stimulated to produce its own antibodies.

Conversely, passive immunity is obtained by receiving antibodies your body did not make itself. This can happen naturally, as when a newborn will receive its mother's antibodies via the placenta and through breastmilk, or it can happen artificially, by receiving a transfusion or injection of antibodies or anti-venom. In both instances, the body receives antibodies it did not make itself.

Type of immunity		Mechanism	Duration of immunity
Active	Natural	After infection with a pathogen, the circulation contains some antibodies and B and T memory cells.	Long-lived immunity, from many years to life
	Artificial	After vaccination with antigens, the circulation contains antibodies and memory cells.	Long-lived immunity, from many years to life
Passive	Natural	Antibodies cross the placenta from mother to the developing foetus. Breast milk also contains some antibodies.	Short-lived immunity: the baby has no plasma cells or memory cells to continue antibody production
	Artificial	At times of high-risk infection, a person may receive an injection of antibody serum. Antitoxins are antibodies specific to toxins. They are given by injection. Antivenins are antibodies specific to the toxins in snake or insect bites.	Short-lived immunity: the antibodies may help a person through a crisis, but the body has no plasma cells or memory cells to continue antibody production

Pathogens fight back

Despite non-specific defences and a well-developed immune system, we still get infectious diseases. This is because pathogens have evolved adaptations that make them less likely to be attacked or killed by the body's defences. See Chapter 4, section 4.1.4 for more information on genetic change within pathogens and challenges for treatment strategies.

- Some populations of pathogens produce antigenic variation by changing their surface antigens as they grow and reproduce inside a host. The body has no immunity to the new antigen.
- Some pathogens develop immunity to digestive enzymes and can therefore survive inside a phagocyte unaffected by its digestive enzymes. Others are able to burrow out of digestive sacs in phagocytes.
- Some bacteria produce spores that are resistant to immune attack.
- Some pathogens produce and live in resistant cysts.
- Some parasites (e.g. malarial *Plasmodium*) live inside cells, avoiding antibody attack.
- Some pathogens attack the immune system itself. They may release enzymes or toxins that destroy leukocytes. For example, the human immunodeficiency virus (HIV), which causes AIDS, attacks helper T cells.

3.3 Disease challenges and strategies

It is certain that almost everyone on the planet has been affected by the disease COVID-19 in some way. Isolation, quarantine and lockdown measures were all strategies put in place to cope with a virus that knows no borders. The challenges faced by the emergence and re-emergence of diseases include antimicrobial resistance, infection control, outbreak containment and vaccine development.

3.3.1 The emergence of pathogens in a globally connected world

The emergence of new diseases or infections can be caused by previously unknown or undetected pathogens; known pathogens whose role in a specific disease was previously unrecognised; **zoonotic** diseases, where diseases previously only occurring in animals are able to pass to humans and cause illness; or when known pathogens spread to new populations or geographic locations. The dramatic impact of emerging diseases on a population was seen during the colonisation of Australia by Europeans.

In the late 18th century, at the time of European arrival, an estimated 320 000 Aboriginal and Torres Strait Islander peoples occupied Australia. The majority lived in the south-east of the country. The landing of the European colonisers introduced human disease such as tuberculosis, influenza, measles, syphilis and smallpox.

The outbreak of these exotic diseases resulted in epidemics with high rates of morbidity (illness) and mortality (death) among Aboriginal and Torres Strait Islander peoples. By the 1930s, a combination of factors including the emergence of diseases and the restriction of Aboriginal and Torres Strait Islander peoples to 'reserves', where they were confined and isolated from European settlers, resulted in a massive population decrease. Exact figures are not known, but it is believed that approximately only 80 000 Indigenous people remained in Australia at this time.

If infection rates rise again for a disease that is thought to have been eradicated, this is known as a re-emerging infectious disease. The re-emergence of tuberculosis is due to the evolution of the bacteria that causes the disease. The bacteria gained resistance to antibiotics, allowing it to proliferate. The *Plasmodium* that causes malaria has also become resistant to treatment, and the mosquito that acts as a vector for the *Plasmodium* has acquired resistance to pesticides. Other examples of re-emerging diseases include diphtheria and whooping cough, which is caused by an insufficient number of vaccinated people within a population.

If vaccination levels are not enough to provide **herd immunity**, a pathogen can be introduced (or reintroduced) to a population and lead to an outbreak. The diagram below illustrates the importance of herd immunity in protecting a population. Herd immunity will be discussed in more detail in Section 3.3.3.

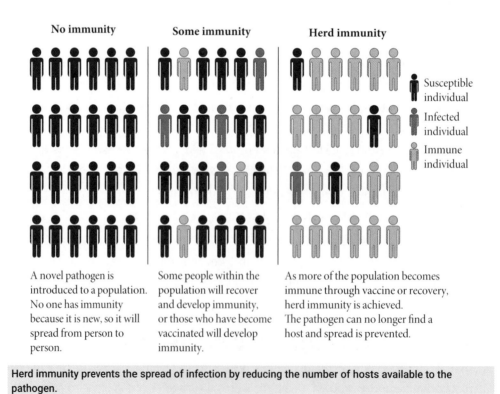

| No immunity | Some immunity | Herd immunity |

Susceptible individual
Infected individual
Immune individual

A novel pathogen is introduced to a population. No one has immunity because it is new, so it will spread from person to person.

Some people within the population will recover and develop immunity, or those who have become vaccinated will develop immunity.

As more of the population becomes immune through vaccine or recovery, herd immunity is achieved. The pathogen can no longer find a host and spread is prevented.

Herd immunity prevents the spread of infection by reducing the number of hosts available to the pathogen.

3.3.2 Controlling the spread of pathogens

Diseases can start locally and spread across the globe quickly. SARS-CoV-2, the virus that causes the disease COVID-19, first emerged in China and, within months, a pandemic was declared. We know a lot about disease-causing organisms because:

- we collect a lot of information about disease outbreaks: regions, numbers, modes of transmission
- the spread of disease can be influenced by human customs and behaviour, as much as by the features of the pathogen (e.g. MERS patients escape from hospital, Ebola continues to spread).

An **infectious disease** is one that is caused by a pathogen such as bacteria, viruses and parasites and can be spread from host to host either through direct or indirect contact or through a medium such as food, water or insect vectors. Examples of infectious diseases are food poisoning, Lyme's disease and urinary tract infection. A **contagious disease** is a subset of infectious diseases in that they are caused by a pathogen and spread from one organism to another through direct or indirect contact. Examples of contagious diseases are flu, chickenpox and COVID-19. A **non-infectious disease** is one that cannot be spread from one host to the next, such as Alzheimer's disease or diabetes.

The following terms are important in understanding infectious diseases.

Important key terms for infectious diseases	
Endemic	A disease is said to be endemic when it is regularly present among a particular group of people or in a particular area; e.g. malaria is endemic to many tropical regions of the world.
Outbreak	An outbreak is the occurrence of one or several cases of a disease in an area in which it is not normally present; e.g. a measles outbreak in a Melbourne suburb where immunisation rates are low.
Epidemic	An epidemic is the widespread occurrence of an infectious disease in a particular group of people in a particular time period; e.g. an epidemic of SARS (severe acute respiratory syndrome) occurred in southern China, including Hong Kong, between November 2002 and July 2003. There were thousands of cases, including a small number in neighbouring countries, and nearly 800 people died.
Pandemic	A pandemic is an epidemic on a global scale. It is called a pandemic when it spreads beyond one region and affects people on multiple continents. HIV is an example of a global pandemic. The Flu pandemic of 1918–19 is estimated to have killed about 50 000 000 people across all continents.
Isolation	Isolation is the separation of people who are already sick from the general population. Isolation often occurs in specifically designed hospital wards or other facilities.
Quarantine	Quarantine is the separation of healthy people or animals known to have been exposed to a pathogen or an infected individual. If they develop symptoms, they are removed to isolation; if they do not develop symptoms, they are allowed to return to normal life after a period of time.

Identification of pathogens

There are many ways in which biological knowledge contributes to our knowledge and identification of pathogens and disease. These can be roughly divided into three categories:

- phenotype – examination of the physical and chemical aspects of the organism or agent
- immunological – use of antibodies to detect antigens, or use of antigens to detect antibodies
- molecular – analysis of molecules, particularly nucleic acids, to identify pathogens and design treatments.

Methods of identification of pathogens and disease	
Phenotype 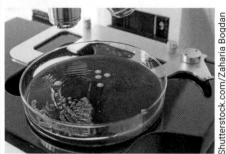 Shutterstock.com/Zaharia Bogdan Rares	Morphology – macroscopic (shape, size or colour of bacterial colonies) and microscopic (isolation and examination of the pathogens using microscopes and staining techniques) morphological features can be used to identify pathogens. Selective media – bacteria grow differently in different media. They can be cultured on agar plates containing indicators that change colour when different bacterial products are formed. Selective media can also be used to test for the susceptibility of the pathogen to a range of drugs.
Immunological 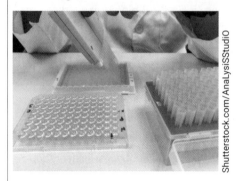 Shutterstock.com/AnalysiSStudiO	Serology – blood (or other fluids) tests look for antibodies produced by the patient. A known antigen may be added to a blood sample and, if the corresponding antibody is present, the two proteins will bind together. The bound antigen–antibody can be detected in a variety of ways. Labs can also use manufactured antibodies to look for the presence of known antigens on bacteria or viruses using techniques such as enzyme-linked immunosorbent assay (ELISA). This technique uses an antibody–enzyme complex. The antibody part of the complex binds with the pathogen antigen. The enzyme's usual substrate is then added and reacts with the enzyme, producing a colour change that can be detected. The variable haemagglutinin and neuraminidase antigens on the surface of influenza viruses can be determined by their reaction to monoclonal antibodies. Antibodies specific to H1, H2, H3, N1 etc. are used to identify and track flu outbreaks.
Molecular 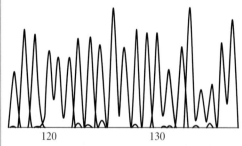 120 130 G A T A A A T C T G G T C T T A T T T C C	The pathogen's nucleic acids can be sequenced using techniques such as PCR and DNA–DNA hybridisation, and comparisons made to data banks of sequenced pathogens. Sequencing of pathogen nucleic acids helps to track mutations that cause antigenic variation and to determine the source and spread of particular pathogen strains. Amino acid sequencing of protein molecules present in infectious diseases has been used to track the spread of plant pathogens in food crops. Protein sequencing is also important in developing antibodies for use in immunological tests. Full genome sequencing has been used to reveal the approximately 13 500 nucleotide sequence of the flu virus genome. Each year, the CDC (Centers for Disease Control and Prevention, in the USA) performs whole or partial genome sequencing on about 7000 influenza viruses collected from various parts of the world, tracking seasonal flu, swine flu and bird flu varieties. The sequence information is also important in the production of new vaccines each year.

Modes of transmission

The way in which pathogens can be passed on from one individual to another is called **transmission**. Research on the life cycles of pathogens, including their hosts and modes of transmission, provides points at which preventive measures can be used to interrupt this cycle and prevent spread.

Mode of transmission	Example of disease	Source of infection	Prevention
Airborne droplets	Measles, influenza, whooping cough	Respiratory droplets	Face masks, good hygiene, avoid bodily fluids
Direct contact	HIV	Bodily fluids	Avoid bodily fluids, good hygiene
Indirect contact	Athlete's foot fungus	Fomites – these are objects such as clothing and furniture that can carry infection	Good hygiene, avoid walking barefoot
Ingestion	Salmonella poisoning	Infected food and beverages	Safe food and water, good hygiene, food safety
Endogenous; originating from within the organism but was previously asymptomatic	Viruses and bacteria	Catheter	Good hygiene
Vectors	Malaria	Mosquito	Good hygiene, immunotherapy

Controlling pathogens

There are treatments to control the development of a disease within an individual. These include:

- Antibiotics – stop the progression of bacterial diseases by damaging or slowing the growth of the disease-causing bacteria.
- Antivirals – medicines to stop viruses from infecting and multiplying in healthy body cells.
- Fungicides – chemical compounds used to kill parasitic fungi and their spores.

There are social strategies that can control the spread of pathogens within a society. These include:

- Improved public health education
- Public health care
- Physical distancing between people
- Face masks
- Hand hygiene
- Vaccination
- Quarantine

3.3.3 Vaccination programs and their role in immunity

After coming into contact with a pathogen (and surviving!), the body develops immunity. This is due to the presence of antibodies in your circulation and the presence of memory cells. If a person is infected by the same pathogen again, memory cells are ready to multiply and again produce a huge quantity of plasma cells and antibodies. The second immune response is faster and bigger than the first. This second infection is unlikely even to produce symptoms because the circulating antibodies, and antibodies produced by a new crop of plasma cells, destroy the pathogen before any symptoms appear. The relative amounts of antibodies produced in the first and subsequent infections is shown in the graph at the right.

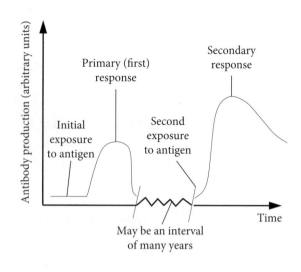

Immunisation

Vaccines are used to induce active immunity. In vaccination, a pathogen's antigens are introduced into a host, and this acts like a first infection with a pathogen. The host's immune cells mount an immune attack and produce memory cells in case of future exposure to the same pathogen.

Vaccines are solutions, usually injected, that contain:

- bacteria or viruses that have been killed (inactivated)
- live attenuated (weakened) bacteria or viruses, which usually retain the ability to live and/or reproduce but can no longer cause disease
- toxoids – chemical copies of bacterial toxins – that have been inactivated so they cannot cause disease
- only part of the microbe – the part that contains the antigens stimulates an immune response; these are subunit vaccines
- conjugates – some bacteria have polysaccharide outer coats that are not readily recognised by the immune system. By linking proteins (e.g. toxins) to these polysaccharide coats, the immune system can be helped to recognise the polysaccharide as if it were a protein antigen. A conjugate vaccine is used in the *Haemophilus influenzae* type B vaccine
- DNA – DNA vaccines are also being developed, although they are still experimental at present. They cause the host's cells to express the viral gene by producing molecules that the virus itself would induce. The immune system recognises them as foreign and launches an immune attack.

Herd immunity

As well as protecting individuals against infection, large-scale immunisation programs provide herd immunity in the community. Fewer potential sufferers means there are fewer hosts available for the pathogen to infect and spread from. Consider the following two school populations exposed to the measles virus.

In population A, only a few students have been immunised (shaded individuals). The virus initially infects one individual. She gets ill and spreads the virus to several susceptible students. They, in turn, spread the virus to several more. Soon, nearly everyone is sick. Only the immunised children escape infection.

In population B, a student who is not immunised brings the virus to school. All his friends are immunised. They breathe in the virus, but rapidly produce antibodies against it and the virus is killed off before it makes them sick and before they are able to spread it. Children in other classes, even if not immunised, have a chance of avoiding the measles as long as they don't come into contact with the first sufferer. They are protected by the immunised 'herd' in their school.

For herd immunity to prevent the spread of disease a high proportion of the population needs to be immune. The exact proportion depends on the virulence and infectivity of a specific disease. For

9780170479431

measles, a highly contagious and virulent disease, 95% of the population must be exposed to the disease or vaccinated for herd immunity, but for polio, it is 80%.

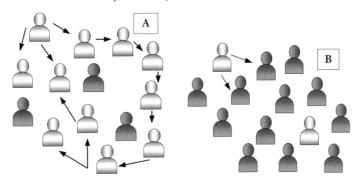

3.3.4 The development of immunotherapy strategies

Monoclonal antibodies

A **monoclonal antibody (MAb)** is a laboratory-produced antibody molecule that binds to one specific part (epitope) of one specific antigen. Monoclonal means that the antibodies are *one* type, produced by *one* cell line, containing cells that are clones (exact copies) of *one* original antibody-producing cell. The following diagrams outline how monoclonal antibodies are produced.

1 A mouse is injected with antigens in order to stimulate an immune response. 	**2** The mouse produces plasma cells and antibodies in response to the foreign antigens. 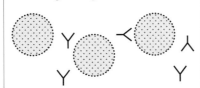	**3** B lymphocytes (plasma cells) from the mouse's spleen are extracted. 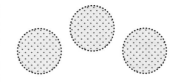
4 B lymphocytes are mixed with cultured myeloma cells in a solution of polyethylene glycol. Some cells fuse, becoming hybrids of the two cell types. 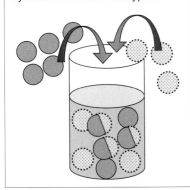	**5** The fused cells contain the genes needed to produce antibodies (from the B cells) and the ability to go on dividing and not die (from the myeloma cells). 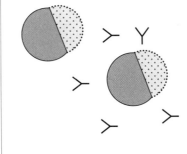	**6** The cells that produce the desired antibody are selected and grown in a mass culture. As they are all the same (clones), they produce the same type of antibody. The antibody can be collected. 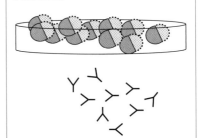

Treatment of autoimmune diseases and cancers

Autoimmune diseases are caused when the body's immune system is unable to differentiate between self and non-self antigens, and produces autoantibodies that attack the body's own cells. Monoclonal antibodies (MAb) can be used to treat a number of diseases, including some forms of arthritis, bone disorders such as osteoporosis, systemic lupus and other autoimmune disorders. The MAb is able to bind to cytokines and block inflammation.

Monoclonal antibodies are a relatively new treatment for some cancers. Antibodies delivered intravenously bind to the specific regions of antigens present on the surface of cancer cells. They affect the cancer cell in the same way that antibodies produced by the body's own immune system can affect cells. These include:

- making the cancer cell more visible to the immune system. A monoclonal antibody can attach to a cancer cell, making it easier for the immune system to find
- blocking receptors on the surface of the cell that normally bind with signalling molecules that cause the cancer cell to grow and multiply. Cancer cells often have more of these receptors than other cells
- stopping new blood vessels from forming. Tumours rely on increased blood supply to grow, so blocking the molecules that attract new blood vessels can starve a tumour
- delivering radiation to cancer cells. By combining a radioactive particle with a monoclonal antibody, radiation can be delivered directly to the cancer cells
- delivering chemotherapy to cancer cells. Combining chemotherapy drugs with a monoclonal antibody means the drug is delivered in a targeted way directly to the cancer cell.

In some cases however, monoclonal antibodies may have no effect on the immune response.

Drug	Target	Effects
Rituximab (Rituxan®)	A specific protein (CD20) found only on B cells, including B cells involved in certain lymphomas	Attaches to the CD20 protein on the B cells; this makes the cells more visible to the immune system, which can then attack
Cetuximab (Erbitux®)	Colon cancer, head and neck cancers	Attaches to receptors on cancer cells that accept a growth signal, blocking this signal from reaching its target on the cancer cells
Bevacizumab (Avastin®)	A growth signal that cancer cells send out to attract new blood vessels	Intercepts a tumour's signals and stops them from attracting blood vessels
Ibritumomab (Zevalin®)	Cancerous blood cells in non-Hodgkin's lymphoma	Attaches to receptors on cancerous cells and delivers a dose of radiation

Glossary

agglutination When antigens or pathogens become stuck together because of antibody binding

allergen A protein that causes an inappropriate allergic immune response

antibody A specific protein that recognises and binds to a specific antigen; circulating antibodies are produced and secreted by plasma B cells; also known as an immunoglobulin

antigen Any molecule, usually a protein, that triggers an immune response, and can be recognised by a specific complementary antibody

apoptosis Regulated cell death, which is necessary to get rid of old, damaged or unnecessary cells

cell-mediated immune response A specific response to non-self eukaryotic cells; aids in defence against cells infected with viruses, cancerous cells and transplanted tissue

clonal expansion The process where many identical daughter cells are produced from a parent cell

clonal selection Random genetic mutations during production allow for a diverse range of lymphocyte receptors to be generated. These are selected for recognition of a complementary antigen by a B lymphocyte or T lymphocyte

complement protein Protein present in body fluids; can be activated to fight infection directly by the presence of pathogens or when antigens and antibodies combine; part of non-specific defences

contagious disease Disease caused by a pathogen and can be spread from host to host through direct or indirect contact

cytokines A large group of proteins, peptides or glycoproteins that are secreted by specific cells of the immune system; they generally affect other cells of the immune system

cytotoxic T cell (T$_C$ cell) A T lymphocyte capable of binding with, and destroying, non-self eukaryotic cells, cancerous cells or cells infected with viruses

death receptors Cell surface membrane receptors which, when activated, trigger a signalling cascade that results in apoptosis

degranulation A cellular process that releases molecules known as granules from secretory vesicles to the extracellular environment

fever A body temperature above the normal range

first line of defence The physical and chemical barriers that keep pathogens from entering the body of a living thing

helper T cell (T$_H$ cell) A lymphocyte that activates other T cells and B lymphocytes

herd immunity A form of immunity where unvaccinated individuals are protected against a disease because a large number of people are vaccinated or are otherwise immune, thereby making it unlikely that unvaccinated people will come in contact with anyone suffering from the disease

histamine A substance released from mast cells, which is a response for inflammation and allergic reactions

humoral immune response A specific response to foreign antigens; involves the production of antibodies by plasma cells, leading to the destruction of pathogens; also called antibody mediated immunity

immunity Having resistance to infection by a specific pathogen

immunoglobulin A specific protein that recognises and binds to a specific antigen; circulating immunoglobulins are produced and secreted by plasma B cells; also known as an antibody

infectious disease Disease caused by a pathogen and can be spread from host to host either through direct or indirect contact or through a medium

inflammation The response of body tissue to injury or infection; causes swelling, redness and heat; the response is due to dilation and leaking of blood vessels and phagocytosis

innate immune response A response to a pathogen that is not specific to the antigen, only that it has been identified as being non-self; the response does not generate antibodies or memory lymphocytes

interferons Chemicals produced by animal cells in response to viral infection; they interfere with viral replication and help cells resist infection

interleukins A group of cytokines that act as messenger molecules between immune cells (inter = between; leukin = white blood cells)

leukocyte A general term for a white blood cell

lymphocyte An agranular white blood cell that is important in the immune response; there are T lymphocytes and B lymphocytes

major histocompatibility complex (MHC) protein markers found on cell surfaces that are important in distinguishing self from non-self; MHC class I is found on all cells and MHC class II is found only on antigen-presenting cells

mast cell A large cell in connective tissue; produces and secretes histamine; important in the inflammatory response and in allergic reactions

memory cell A cell produced by B or T lymphocytes in response to recognition of a foreign antigen; remains in the circulatory system for many years

monoclonal antibody (MAb) A laboratory-produced antibody molecule that binds to one specific part of one specific antigen

natural killer (NK) cell a circulating leukocyte that kills body cells infected with a virus or transformed by cancer

neutralisation The process by which antibodies prevent toxins from acting by binding to them and blocking them from binding to their targets

non-infectious disease Disease that cannot be spread from host to host

non-self antigen A molecule that is not recognised by the immune system as being part of the organism itself

opsonisation The process where a pathogen is targeted for destruction by the immune system

pathogen An organism foreign to the body and capable of causing disease

perforin A glycoprotein that is able to form pores in the cell membrane of other cells

phagocyte A leucocyte that engulfs, ingests and digests foreign matter such as pathogens, debris or damaged cells

phagocytosis A process by which phagocytes engulf a particle or cell

plasma B cell A B lymphocyte that produces and secretes large quantities of antibodies

prostaglandins A group of lipids that control processes such as inflammation, blood flow and fever; e.g. pyrogens are prostaglandins

pyrogen A prostaglandin that acts on the hypothalamus, trigging the body to generate heat resulting in a fever

self antigen An antigen or a molecule that is a normal body component

transmission The passing of an infectious disease from an infected host to another individual or group

vaccination Deliberate introduction of weakened or dead foreign antigens into the body so that the body mounts an immune response that results in immunity should the body be re-exposed

zoonotic Describes a disease that can be transmitted from animals to humans

Revision summary

In the table below, provide a brief definition of each of the key concepts. This will ensure that you have revised all the key knowledge in this Area of Study in preparation for the exam.

How do organisms respond to pathogens?	
Physical, chemical and microbiota barriers in animals and plants	
The innate immune response, including the steps in an inflammatory response	
Roles of macrophages, neutrophils, dendritic cells, eosinophils, natural killer cells, mast cells, complement proteins and interferons	
Initiation of an immune response, including antigen presentation	
The distinction between self antigens and non-self antigens	
Cellular and non-cellular pathogens and allergens	
The role of the lymphatic system and lymph nodes as the sites for antigen recognition by T and B lymphocytes	
Clonal selection and clonal expansion	
Adaptive immune response against extracellular threats, including the actions of B lymphocytes and their antibodies and helper T cells	
Adaptive immune response against intracellular threats including actions of helper T and cytotoxic T cells	
Natural vs artificial immunity; active vs passive strategies for acquiring immunity	
New pathogens and re-emergence of known pathogens, including the impact of European arrival on Indigenous peoples	
The distinction between infectious and contagious diseases	
Scientific and social strategies to identify and control the spread of pathogens, including host, mode of transmission and measures to control transmission	
Vaccination programs and their role in maintaining herd immunity for a disease	
Immunotherapy strategies, including monoclonal antibodies for treatment of autoimmune disease in a human population	

Exam practice

Responding to antigens

Solutions start on page 231.

Multiple-choice questions

Question 1 ▣▢▢

Which of the following prevents pathogens from entering the human body?

A Intact skin

B The normal flora that colonises the surfaces of humans

C Macrophages that quickly engulf pathogens

D Stomach acid

Question 2 ▣▢▢

Which one of the following is an example of a plant defence against a pathogen?

A Marigolds excreting chemicals into the soil that are toxic to nematodes

B Macrophages migrating to a leaf that has been infected by a fungus

C The production of antibodies that circulate around the plant leaves and stems

D Waxy leaf surfaces acting as a chemical barrier

Question 3 ▣▣▢

Frogs can become infected with a pathogenic fungus called *Batrachochytrium dendrobatidis* (*Bd*). Scientists have found that fungal infections are significantly reduced in the warm, wet season and increase in the dry, cool season. They also found that in the warm, wet season the skin of the frogs remained saturated with moisture and a large number of beneficial bacteria. In the dry, cool season, their skin still contained the diversity of beneficial bacteria but in lower numbers, and their skin was only moist.

What could be an explanation for the low numbers of infection in the warm, wet season?

A The large numbers of beneficial bacteria help to prevent the fungal spores from attaching to the frog's skin, so the fungus cannot obtain enough nutrients to survive due to the competing bacteria.

B The fungus was able to attach and begin to grow but the warm, wet environment resulted in the developing fungus rotting due to all the bacteria, and it died before being able to cause infection.

C The pathogenic fungus preferred the cool, dry season, so was not in its optimum temperature range to cause disease in the frog.

D In the warm, wet season, the continual movement of the frog limits the number of fungal spores attaching to the frog's skin, as they are quickly washed away while the frog is swimming.

Question 4 ▣▣▢

If an airborne virus gained entry into the bloodstream, the body would mount a response to the pathogen. Non-specific defences involved would include the

A inflammatory response that is part of the T cell defence.

B release of histamine from macrophages.

C action of platelets adhering to the virus.

D release of interferon to disable virus-infected cells.

9780170479431

Question 5 🔘🔘⚪

Dendritic cells

A are part of the body's specific immune response only.

B are produced by the thymus.

C present non-self antigens to lymphocytes.

D are involved in the inflammatory response.

Question 6 ©VCAA VCAA 2015 SA Q18 🔘🔘⚪

A girl is carrying a piece of wood. A small piece breaks off and becomes embedded in her finger. The next day, she notices an inflammatory response occurring in her finger. In the region around the small piece of wood embedded in her finger

A mast cells would release antibodies.

B the skin tissue would become pale and cold.

C the capillaries would become more permeable.

D red blood cells would leave the blood vessels and engulf foreign material.

Question 7 🔘🔘🔘

William travelled overseas and began to feel ill on his return home. His doctor diagnosed a hookworm infection. Hookworms are an intestinal parasite that feed on blood obtained through the wall of the intestines they latch (hook) onto with their mouth parts. The cell type that is part of the innate immune response that will be increased in William's blood stream is

A eosinophils.   C memory B cells.

B helper T cells.   D dendritic cells.

Question 8 🔘⚪⚪

Molecules on the surface of cells that enable the cells to distinguish self from non-self cells are known as

A antibodies.   C antigens.

B membrane receptors.   D lipoproteins.

Question 9 🔘⚪⚪

An adult human female has around 200 different cell types. An example of 'self' material would be

A *E. coli* bacteria in her gastrointestinal tract.

B influenza virus that is infecting her respiratory tract.

C sperm present in her reproductive tract.

D the epithelial cells lining the inside of her ears.

Question 10 ©VCAA VCAA 2014 SA Q17 ●●○

The following diagrams represent various types of plant and mammal pathogens. Their approximate size is indicated by a scale bar.

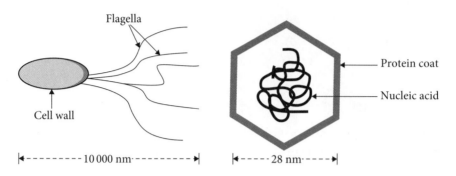

Pathogen W   **Pathogen X**

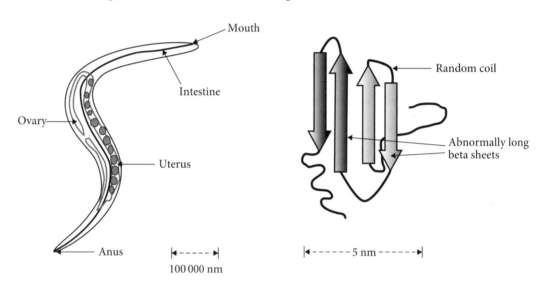

Pathogen Y   **Pathogen Z**

With respect to these pathogens, it would be reasonable to state that

A Z is a virus.

B both W and X are cellular.

C Y reproduces by binary fission.

D X reproduces by invading host-cell DNA.

Short-answer questions

Question 11 (4 marks) ●○○

Name and describe **two** different types of barriers that pathogens might encounter
as they enter the human respiratory tract.   4 marks

Question 12 (4 marks) ●●○

Photosynthesising plants store energy from the Sun and include vitamins and minerals beneficial in the diets of many different organisms. Plants and animals have very different immune systems. While animals have a complex system of specific and non-specific mobile immune cells and chemicals, plants have many barriers that prevent the entry of pathogens.

a Describe **one** chemical barrier and one physical barrier in a plant that can protect it
from invading pathogens.   2 marks

b The human immune system has very different defence barriers. Describe **one** physical
barrier and **one** chemical barrier that prevent pathogens from entering the human body.   2 marks

Question 13 (6 marks) ▣▣▢

Copy and complete the table below, describing the function of each of the cells and where they can be found, including any chemicals they produce in the innate immune response.

Cell type	Function
Macrophage	
Neutrophil	
Dendritic cell	
Eosinophils	
Natural killer cells	
Mast cells	

Question 14 (9 marks) ▣▣▣

The skin normally provides an effective barrier to the entry of pathogens. At times, however, this barrier is breached, and foreign material can enter the deeper tissue. The site of entry of foreign material, such as bacteria, into the skin eventually appears red, swollen and hot, and may contain pus.

a What name is given to the body's response to the entry of foreign material, outlined above?   1 mark

b Explain how the action of histamine contributes to the appearance of the area after infection.   3 marks

c Outline **two** ways in which the blood's complement proteins may assist in the destruction of the invading bacteria.   2 marks

d Why does complement have very little effect during viral infections?   1 mark

e Following an inflammatory response, lymphocytes in the lymph nodes may become activated. Outline the mechanism that leads to the activation of helper T lymphocytes, despite them not being at the site of the infection.   2 marks

Question 15 (7 marks) ©VCAA VCAA 2019 SB Q3 ▣▣▣

The human immune system consists of a series of defensive barriers that protect the body from infection. When bacteria come into contact with the body, they immediately encounter these defences and must bypass each barrier if they are to survive and infect the body.

a When bacteria come into contact with the body, they must gain access to the living tissues to become pathogens. List **two** possible routes the bacteria could use to access the living tissues of the body.   2 marks

b Once bacteria are within or have access to the living tissues of the body, but before cells are aware of their presence, the bacteria will encounter chemical barriers. Name **one** of these chemical barriers and explain its function.   2 marks

c When an inflammatory response starts, the first cellular responders will be cells from the innate immune system. One of these cells releases histamine. How does histamine contribute to the inflammatory response?   1 mark

d If bacteria are not destroyed by innate immune responses, adaptive immune responses become involved. Describe how an adaptive immune response is initiated during a bacterial infection.   2 marks

Question 16 (9 marks) ▣▣▣

Hay fever is the general term given to an allergic reaction to airborne substances such as pollen. The symptoms of an allergic reaction are swelling, redness, sneezing, runny nose and contraction of smooth muscle, which can make breathing difficult.

a What name is given to a substance that causes an allergic reaction?   1 mark

b Why are conditions such as hay fever only suffered at certain times of the year?   1 mark

c **i** What chemical produced by mast cells is responsible for the symptoms of allergies?   1 mark

    ii Explain how the substance in part **ci** is able to bring about the swelling
and redness so evident in hay fever.   2 marks

d Explain how the mast cells of some individuals are able to respond very rapidly
to the presence of pollen.   2 marks

e List **two** strategies that a person could take to reduce their risk of developing hay fever.   2 marks

Question 17 (6 marks) ◐◐○

There are two types of MHC proteins embedded in cell membranes. MHC class I are found on all
nucleated body cells and MHC class II are found only on antigen-presenting cells.

a Name the type of MHC protein found on the following cells.

    i Skin cell   1 mark

    ii Macrophage   1 mark

    iii Red blood cell   1 mark

b MHC markers present antigens to be identified by the adaptive immune system. Describe
the roles of the **two** different MHC proteins and why antigens are presented in this way.   3 marks

Acquiring immunity

Solutions start on page 235.

Multiple-choice questions

Question 1 ○○○

Which of the options below correctly identifies the link between the innate immune response and the
specific immune response?

A Phagocytes present parts of engulfed pathogens on MHC class I antigens to cytotoxic T cells in the
lymph nodes.

B Dendritic cells present antigens from engulfed pathogens on MHC class II molecules to helper T cells
in the lymphatic system.

C Neutrophils present entire pathogens to B cells in order to produce specific antibodies at the site of
infection.

D Macrophages present antigens to helper T cells in the bloodstream to recruit the assistance of other
immune system cells.

Question 2 ©VCAA VCAA 2017 SA Q23 ◐◐○

The lymphatic system includes the lymph nodes, spleen and tonsils.
In these particular organs

A clotting factors are inactivated to help seal a wound.

B clonal selection and proliferation of B cells occurs.

C non-self antigens are identified by red blood cells.

D the initial response to an allergen is triggered.

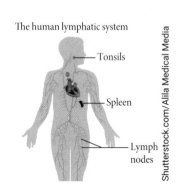

The human lymphatic system

Tonsils

Spleen

Lymph
nodes

Shutterstock.com/Alila Medical Media

Question 3

The most common surgical treatment for a thymus tumour is the complete removal of the thymus gland. What might be a side effect on the immune system from this surgical procedure?

A Overproduction of mature B cells

B Reduced production of mature T cells

C Complete lack of immune system cells to fight pathogens

D Overproduction of mature B and T cells

Question 4

Cytotoxic T cells

A produce antibodies that will bind to pathogens entering the body.

B are part of the non-specific immune response.

C are able to kill cells infected by viruses.

D engulf pathogens and present fragments to other T cells.

Question 5

When the adaptive immune response is triggered, plasma B cells produce high numbers of proteins called antibodies. In order for these cells to be efficient at producing large numbers of antibodies to combat an infection, plasma B cells must have

A a very large nucleus that takes up most of the cell.   C lots of ribosomes.

B only a few mitochondria.   D a small Golgi apparatus.

Question 6

Helper T cells are activated by antigen-presenting cells. An antigen-presenting cell displays the antigen on MHC molecules for helper T cells to recognise and initiate the specific immune response. An example of an antigen-presenting cell might be a

A helper T cell.   C plasma cell.

B dendritic cell.   D mast cell.

Question 7 ©VCAA VCAA 2015 SA Q16

Consider the following diagram of four pathogens and three antibodies. Which one of the following statements is correct?

Pathogen

R

S

T

U

Antibody

E

F

G

A Antibody E would be effective against both pathogen S and pathogen R.

B Antibody F is effective against three of the pathogens.

C There are no antibodies effective against pathogen U.

D Antibody G is only effective against pathogen R.

Question 8 ○●●

In the humoral immune response

A T cells undergo clonal selection and expansion once an antigen has been recognised.

B helper T cells secrete powerful toxins that destroy bacteria.

C plasma B cells assist T cells to proliferate and produce cytokines that support phagocytosis.

D some of the B cells become memory cells that can rapidly divide and form plasma cells upon exposure to the pathogen a second time.

Question 9 ●●●

Which of the following is an example of active immunity?

A Receiving an injection of antibodies after exposure to a pathogen

B A cancer patient being treated with monoclonal antibodies

C Production of antibodies and memory cells after exposure to a pathogen

D The action of phagocytes in an inflammatory response

Short-answer questions

Question 10 (9 marks) ●●●

a Briefly describe the lymphatic system.   2 marks

b Name **three** bodily structures or organs that are included in this system and describe their functions.   3 marks

c List the primary lymphoid organs and the secondary lymphoid organs.   2 marks

d Does the fluid in the lymphatic system get pumped around the same as blood? Explain.   1 mark

e Why is the lymphatic system important in the immune response?   1 mark

Question 11 (8 marks) ●●●

The lymphatic system is a network of tissues and organs that is crucial for the functioning of the immune system. Lymph nodes are part of this system and are found at various locations around the body, as shown in the diagram at the right.

a What is the function of lymph nodes and why are lymph nodes found in these locations?   3 marks

b Name and describe the function of a cell that is a part of the innate immune response that can be found in lymph nodes.   2 marks

c **i** Describe a cell that is part of the humoral immune response that can be found in lymph nodes.   1 mark

    ii What happens to the cell after it is activated?   2 marks

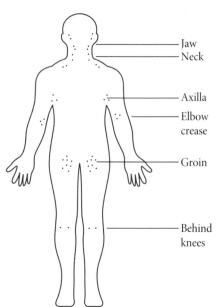

Question 12 (8 marks)

In 1994, there was an outbreak of an illness in a large racing horse stable near Brisbane. It was found to be caused by a *Henipavirus* named Hendra virus. The outbreak involved 21 stabled racehorses and two human cases who contracted the virus through close contact while caring for ill horses. It was found that the virus is transmitted through exposure to body fluids from an infected horse.

a People who become infected with Hendra virus first show symptoms of fever, headache, fatigue and increased respiratory and nasal mucous. Describe what is happening in the body to cause the appearance of these symptoms, after the virus has broken through the initial lines of defence.   4 marks

b Once detected by the human immune system, outline the process that occurs for the body to mount a cell-mediated immune response and memory of the infection.   4 marks

Question 13 (8 marks)

Indicate whether the following statements are true or false.

a Antibodies are effective at eliminating viruses.

b B cells produce pathogen-specific antibodies.

c Antibodies circulate through the lymphatic system in high numbers, providing lifelong immunity.

d Complement proteins are effective against bacterial pathogens but not viral pathogens.

e Only B cells undergo clonal selection and clonal expansion.

f Both B and T lymphocytes create memory cells that are stored in the lymph nodes.

g Helper T cells produce cytokines that stimulate B cell proliferation and clonal selection.

h Cytotoxic T cells engulf pathogens.

Question 14 (10 marks)

a What immune cells produce antibodies?   1 mark

b Draw the structure of an antibody, highlighting the variable region, antigen binding site and light and heavy chains including bonding.   3 marks

c What is the most common type of antibody produced?   1 mark

d What type of antibody is the first to be produced upon initiation of the specific immune response?   1 mark

e Name the **four** roles antibodies play in the immune response.   4 marks

Question 15 (4 marks)

Identify whether the following situations are active or passive, and natural or artificial immunity.

a Glen got bitten by a snake and required antivenin treatment in hospital.   1 mark

b An infant is breastfed by her mother.   1 mark

c Lucy is recovering from the flu.   1 mark

d Julia is an infant who has just had her 6-month vaccination.   1 mark

Disease challenges and strategies

Solutions start on page 238.

Multiple-choice questions

Question 1

A zoonosis is the term for a disease that

A   emerges from zoos due to the number of different animals on the premises.

B   can be passed from humans to animals.

C   can be passed from animals to humans.

D   can be passed within a species from mother to child.

Question 2

Which of the following situations is NOT cause for a disease to re-emerge within a community?

A   Drug resistant strains of pathogens

B   Immigration of people from developing countries with limited healthcare

C   Insufficient vaccination rates within a community

D   Travellers from developed nations visiting developing countries for philanthropic purposes

Use the following information to answer Questions 3 and 4.

The table below compares how eight diseases spread and the number of people likely to be infected by one other infected person.

Disease	Measles	Whooping cough	Rubella	Polio	Smallpox	Mumps	Severe acute respiratory syndrome (SARS)	Ebola
How it spreads	Airborne droplets	Airborne droplets	Airborne droplets	Faecal–oral route	Airborne droplets	Airborne droplets	Airborne droplets	Bodily fluids
Number of people infected from one other person	12 to 18	12 to 17	6 to 7	5 to 7	5 to 7	4 to 7	2 to 4	1 to 4

Question 3 ©VCAA VCAA 2018 SA Q32

What would be the most effective method of preventing the spread of measles during an outbreak?

A   Wash hands thoroughly after going to the toilet.

B   Establish a 'clean needle' exchange program.

C   Vaccinate all infected people.

D   Isolate all infected people.

Question 4 ©VCAA VCAA 2018 SA Q33

Based on the information in the table, which one of the following statements is correct?

A Ebola is the most contagious disease.

B Polio and smallpox have a similar infection rate.

C More people would die from measles than any other disease shown.

D The faecal–oral route is the most effective means of spreading pathogens.

Question 5

Polio, an infectious viral disease almost eradicated worldwide, continues to be a problem in Pakistan. In Pakistan, several hundred cases are reported every year, but few cases are reported in neighbouring countries where immunisation rates are high. The presence of polio in Pakistan is described as

A an epidemic.

B a pandemic.

C an outbreak.

D endemic.

Use the following information to answer Questions 6–8.

Malaria is a parasitic infection caused by the parasite *Plasmodium*. When a mosquito bites an infected person, the mosquito then becomes infected with the parasite, which has no effect on the mosquito. The infected mosquito can transmit the *Plasmodium* to non-infected individuals they subsequently bite. The parasite travels via the bloodstream to the liver where it matures. Once matured, *Plasmodium* leaves the liver and infects the red blood cells of the individual, resulting in a wide range of symptoms including fever, chills, vomiting, muscle pains and diarrhoea. Nearly half the people in the world live in areas at risk of malaria transmission and it kills almost half a million people each year. But the disease is preventable and treatable.

Question 6

Using the information given, it can be concluded that

A the mosquito is a parasitic vector.

B the parasite only infects mosquitos.

C mosquitos can pass the parasite to one another to increase transmission rates.

D malaria is a contagious disease.

Question 7

The most effective way to reduce the incidence of malaria would be to

A prevent humans from living in malaria-prone areas.

B use a broad spectrum of insecticides to eliminate mosquitos.

C increase spending on drugs to treat malaria such as chloroquine.

D prevent mosquitos from biting humans by using screens on windows and doors as well as personal insect repellent.

Question 8

Why is malaria still a problem when it is preventable and treatable?

A *Plasmodium* keeps changing its surface proteins and so evades the human immune system.

B Mosquitos travel long distances and can infect people in different areas previously free of malaria.

C A large proportion of countries at risk of malaria are developing nations or isolated communities.

D Travellers to areas at risk of malaria become infected and take the malarial parasite home with them, infecting people there.

Question 9 🔘

A family returning to Melbourne from Afghanistan indicates to border officials that they were in a village where three children had polio. All members of the family are well but have not been immunised against polio. An appropriate response by officials would be

A to place the family in quarantine.

B that no response is needed, as the family is well.

C to treat the family with antibiotics.

D to return the family to Afghanistan.

Question 10 ©VCAA VCAA 2017 SA Q26 ⚫⚫

A daily blood sample was obtained from an individual who received a single vaccination against a particular strain of the influenza virus. The individual had no prior exposure to this strain of influenza. The graph below shows the concentration of antibodies present in the individual's blood for this strain of influenza over a period of 65 days.

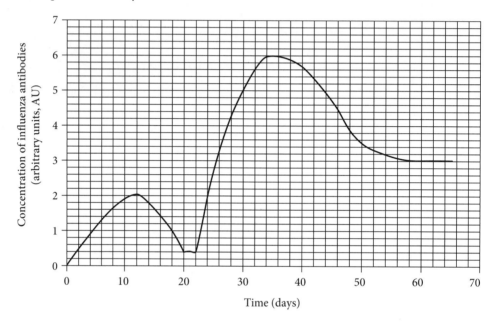

Which one of the following conclusions can be made using this data?

A Memory B cells were activated by exposure to the same strain of the influenza virus on day 22.

B B plasma cells specific to this strain of influenza were most numerous on day 12.

C Herd immunity to this particular strain of influenza was achieved by day 55.

D The vaccination containing weakened influenza antigens occurred on day 10.

Question 11 ©VCAA VCAA 2018 SA Q24 ⚫⚫

Monoclonal antibodies can be produced and used to treat different types of cancers. Which one of the following is a correct statement about monoclonal antibodies?

A Monoclonal antibodies are carbohydrate molecules.

B Monoclonal antibodies produced from the same clone of a cell are specific to the same antigen.

C Monoclonal antibodies pass through the plasma membrane of a cancer cell and attach to an antigen within the cell.

D Monoclonal antibodies produced to treat stomach cancer will be identical to monoclonal antibodies produced to treat breast cancer.

Question 12 ⬤⬤⬤

Monoclonal antibodies can be an effective cancer treatment because they

A bind irreversibly with the DNA of cancer cells.

B bind with surface molecules on cancer cells, making them more visible to immune cells.

C trigger the production of the same antibodies by the patient's own immune system.

D initiate the production of pores within the cell membrane, resulting in the lysis of the cell.

Question 13 ⬤⬤⬤

Which of the following tests could help diagnose an infectious disease in a patient?

A The use of monoclonal antibodies to detect the presence of self antigens in the patient's blood

B The use of synthetic antigens to detect the presence of pathogen antigens in the patient's blood

C The use of monoclonal antibodies to detect the presence of pathogen antigens in the patient's blood

D Comparison of the patient's antibodies with those produced by the pathogen

Short-answer questions

Question 14 (2 marks) ⬤◯

When Europeans started settling in Australia, Aboriginal and Torres Strait Islander peoples started to become ill and die in large numbers. Why might this have occurred?   2 marks

Question 15 (7 marks) ⬤⬤⬤

Tuberculosis (TB) is a disease caused by the bacterium *Mycobacterium tuberculosis*. Although it is relatively rare in Australia today, there were serious outbreaks of the disease up until the 1960s.

M. tuberculosis usually invades and multiplies in lung tissue. Symptoms include fatigue and severe coughing. The lungs respond by producing protective cells that engulf collections of bacteria, forming tubercles. These tubercles enlarge, become crumbly and break away from the lung tissue.

Although the bacteria do not produce toxins as many other pathogens do, they do produce tuberculin, a toxic substance, as dead bacteria disintegrate.

a Name **two** structural features that would identify *M. tuberculosis* as a bacterium.   2 marks

b What is the most likely mode of transmission with this bacterium? Explain your answer.   2 marks

c Routine immunisation against TB occurs in many countries and is available to at-risk groups in Australia. Prior to giving the vaccine, a Mantoux test is given. This involves the injection of a small amount of the toxic substance tuberculin into the skin of the forearm. A positive reaction to the tuberculin causes a raised red patch on the skin. If the Mantoux test is negative, the patient is immunised with an attenuated live vaccine. Why would people previously exposed to the TB bacterium have a positive Mantoux test?   1 mark

d At the beginning of the 20th century tuberculosis was a major cause of death in Australia. A major government strategy resulted in the reduction of tuberculosis numbers across the country with mass screening programs and the development and treatment of Australians with a vaccine. Unfortunately, over the last 10 years Australia has seen the re-emergence of TB. Give **two** reasons why this might be the case.   2 marks

Question 16 (9 marks) ©VCAA VCAA 2019 SB Q9 ⬤⬤⬤

Zika fever is a rapidly emerging viral disease. It is most commonly transferred from one person to another by the *Aedes* species of mosquito. Zika fever in people was discovered in Uganda in 1947. It was thought that a bite from a mosquito had transferred the virus from monkeys to humans. The symptoms of Zika fever are usually mild and 80% of infected humans do not show symptoms. Infection of pregnant women, however, can cause severe defects in their babies.

a One way that diseases, such as Zika fever, are thought to occur is when a pathogen infects humans from an animal host. Identify **one** social or economic factor that could lead to this transfer between hosts.   1 mark

b When scientists attempt to identify a particular disease, they can look for specific antibodies in infected humans. Scientists trying to identify Zika fever infections found that testing for the antibodies produced against the Zika virus often gave them incorrect results. This was because the antibody tests that had been developed could not always identify the difference between the antibodies produced against the Zika virus and the antibodies produced against other viruses. Explain why making a correct identification of a viral pathogen is important in the control of a disease.   3 marks

c Explain why the antibody tests could not identify the difference between the antibodies produced against the Zika virus and the antibodies produced against other viruses. In your response, refer to the structure of the antibody.   2 marks

d *Aedes* mosquitos are not found on every continent. They cannot fly great distances. Vaccines are currently being trialled for the Zika virus. Describe **three** different approaches, other than vaccination, that government health officials could use to reduce the spread of the Zika virus.   3 marks

Question 17 (8 marks) ⬤⬤⬤

Measles is a viral infection caused by a paramyxovirus. It results in respiratory tract inflammation, including sneezing and coughing, a rash and a high fever. It is often accompanied by secondary bacterial infection of the respiratory tract. The virus is usually spread by droplet infection. Antibiotics are often prescribed to minimise secondary infection.

a Infection with the measles virus triggers an immune response. Outline the response of cytotoxic T lymphocytes to viral infection.   2 marks

b Measles is now quite rare. Children are routinely immunised at 12 months of age with a live attenuated vaccine. What is a live attenuated vaccine and why are they used?   2 marks

c The graph right side shows the level of antibody produced by an infant in response to the measles vaccine.

Some years after vaccination, the child was exposed to the measles virus.

  i On the graph, show the response you would expect from this second exposure.   1 mark

  ii Explain why this response occurs.   1 mark

d Mumps is also caused by a paramyxovirus. Would immunisation against measles also give protection against mumps? Explain.   2 marks

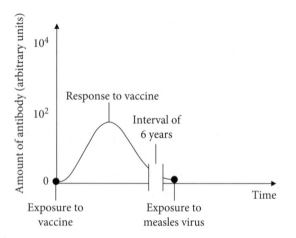

Question 18 (5 marks) ©VCAA VCAA 2018 SB Q5 ⬤⬤

Measles is a highly infectious and dangerous disease. Young children and individuals with impaired immunity are especially susceptible to measles. Analyse the following graphs that show the number of people in the United States of America (USA) who were infected with measles during the period 1954–2000 and the number of people who died as a result of having measles during the same period.

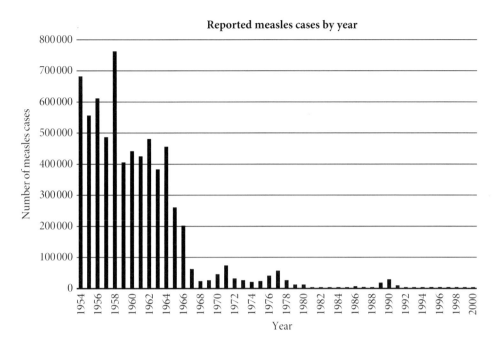

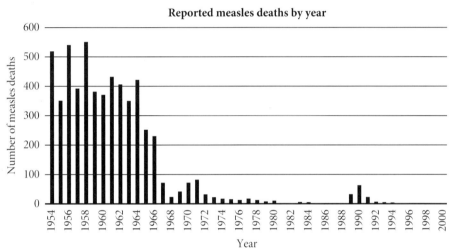

a **i** Which year had the greatest number of reported measles cases? 1 mark

   ii What trends can be observed when the two graphs are compared? 2 marks

b Controlling the number of measles cases in a population relies on herd immunity. What is herd immunity and how does it help control the number of cases of this disease? 2 marks

Question 19 (7 marks)

Scleroderma is a rare autoimmune disease affecting blood vessels and connective tissue. Many internal organs are affected, particularly the oesophagus and kidneys. The skin is also affected, particularly on the face and fingers.

a What is meant by the term 'autoimmune disease'? 1 mark

b What mechanism has broken down in a person who suffers from an autoimmune disease? 1 mark

c Treatment of severe cases of scleroderma involves the use of immunosuppressive drugs, which suppress the function of T and B lymphocytes.

   i Of what advantage is the use of immunosuppressives in the treatment of autoimmune diseases? 1 mark

   ii What are the likely negative side effects of using these drugs? 2 marks

d Monoclonal antibodies can now be used for the treatment of autoimmune disease. What is an advantage of using monoclonal antibodies over the use of immunosuppressant drugs? 2 marks

Chapter 4 Area of Study 2
How are species related over time?

Area of Study summary

This area of study examines variation and adaptations of species and the mechanisms that contribute to evolution. Understanding how to interpret evidence of fossil records, genomic and proteomic comparisons as well as morphological homology leads to the development and continued refinement and review of theories of evolution as new evidence emerges.

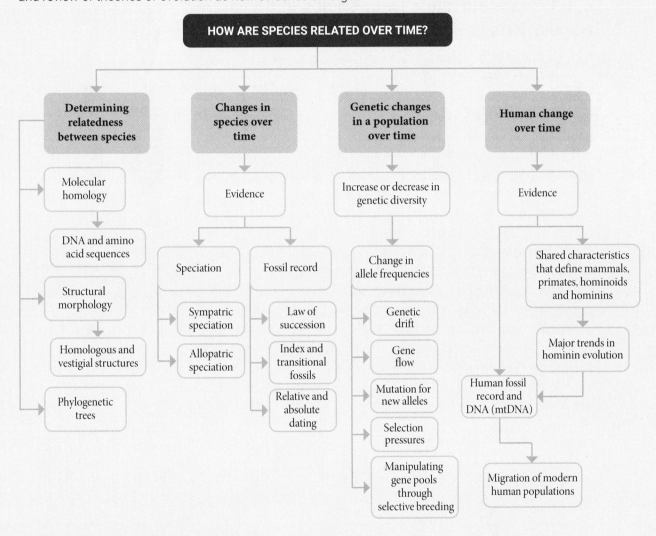

Area of Study 2 Outcome 2

On completing this outcome, you should be able to:

- **understand** how allele frequencies can change within a population
- **apply** knowledge of changing allele frequencies to link to natural selection and speciation
- **analyse** morphological and genetic evidence to support evolutionary relationships
- **evaluate** evidence to determine evolutionary relationships between species
- **create** arguments for evolutionary relationships based on evidence, which can change and adapt as new evidence is discovered.

The key science skills demonstrated in this outcome are:
- **analyse**, **evaluate** and **communicate** scientific ideas
- **create** evidence-based arguments and draw conclusions.
- **analyse** and **evaluate** data, methods and scientific models.

4.1 Genetic changes in a population over time

4.1.1 Changing allele frequencies in a population's gene pool

Members of a group or **species** that interbreed are known collectively as a population. Variation is naturally occurring among populations because of the alternative forms of **alleles** among the genes. The sum total of all of the alleles within a population is known as the **gene pool**. Many genes have more than one alternative allele. For example, domestic dogs can have five possible alleles for coat colour at one gene locus. Any one dog has only one or two of these alleles, but the population contains dogs with 15 different genotypes.

Some alleles are more common in a population than others. There are several factors for changing allele frequencies within a gene pool.

Allele frequencies refer to the proportion of alleles of each type at a particular gene position (locus) in a gene pool. If there are 2 alleles for a gene that codes for a particular characteristic (eg. D and d), their frequencies could be calculated by totalling the number of a particular allele (eg. D) and dividing it by the total number of alleles (d + D) in the population. The allele frequencies are usually expressed as percentages or fractions. If you examine the figure below, the number of alleles of D in the original population is 7/18 = 39% and the allele frequency of d in the original population is 11/18 = 61%. After 20 generations the allele frequency of D is 8/18 = 44% and d is 10/18 = 56%.

Gene alleles

A | a Gene locus

A pair of homologous chromosomes carry the same genes in the same locations; however, there can be different versions of the gene. These versions are known as alleles.

| Original population | After 10 generations | After 20 generations |

Rr BB DD	Rr bb Dd	Rr bb Dd	Rr Bb Dd	RR bb dd	Rr bb dd	rr bb Dd	Rr bb Dd	Rr bb dd
RR bb Dd	Rr Bb dd	rr bb Dd	rr bb dd	Rr bb Dd	Rr bb DD	RR bb Dd	RR bb Dd	RR bb DD
Rr bb dd	rr bb dd	rr Bb Dd	rr bb Dd	RR bb Dd	rr bb Dd	rr bb dd	rr bb DD	rr bb dd

— Time →

Alleles: R = round, r = square; B = blue, b = white; D = broad, d = narrow

Allele frequencies of alleles D and d have changed after 20 generations.

Selection pressures

Allele frequencies can be changed by **natural selection**. Some phenotypes are better suited to particular environments, so organisms that have the alleles for these phenotypes are better able to survive.

On a small scale (**microevolution**), changes occur in allele frequencies and phenotype frequencies within populations and between different populations of the same species. On a larger scale (**macroevolution**), new species arise, old species become extinct, and whole new phyla of organisms appear and change over time. The main force driving evolution is natural selection. The theory of evolution by natural selection was proposed by Charles Darwin and Alfred Russel Wallace in the 1850s.

Darwin's theory of evolution by natural selection was based on his observations of life forms in many parts of the world and by what he inferred from these observations. The theory can best be understood by the following four principles.

1 **Variation** exists between members of species; members of a population are not identical to each other.

2 Organisms produce many more offspring than are needed to replace themselves: plants produce thousands of seeds; animals produce numerous offspring. Overproduction continues despite limited resources and living space. Population sizes, however, remain fairly constant (there is a 'struggle for survival').

3 Some of the variation between organisms makes particular individuals more likely than others to survive and reproduce. These favourable traits are selected for while detrimental or less favourable traits are selected against in the 'struggle for survival'.

4 At least some of this variation is inherited. Organisms that survive to reproduce pass on their favourable traits to their offspring in the process of **inheritance**. Advantageous traits will increase, and unfavourable traits will decrease in allele frequency.

Darwin's theory has often been called the 'survival of the fittest', a phrase that has led to misunderstanding the principle of natural selection. 'Fittest' does *not* mean fastest, strongest, meanest, most aggressive etc. The organisms that survive the struggle to live and reproduce are those whose phenotype gives them some **selective advantage** in their environment. Environments provide **selection pressures** or **selecting agents**: factors that act to favour one phenotype over another.

- A selecting agent may be an aspect of the physical environment (abiotic) such as climate, chemical composition or moisture levels.

- The selecting agent may be the influence of another species, such as predation or parasitism.

- It may be population pressure and competition from members of the same species.

	Antibiotic resistance in bacteria	Peppered moths	Tuskless elephants
Variation	Some bacteria are naturally resistant to antibiotics; others are susceptible. Differences are due to differences in the bacteria's DNA.	Some moths have genes for dark pigmentation and some do not. There are both light and dark moths in the population.	Tusk presence/absence and tusk size are under genetic control and vary in populations of African elephants.
Selecting agent	Presence of an antibiotic	Differential predation by birds based on camouflage	Poaching (illegal killing) of elephants to harvest their tusks
Selective advantage	Bacteria that have the resistant gene to the antibiotic survive. Other bacteria die.	In forests with light-coloured trees, the darker moths are more easily seen and therefore eaten by birds. Light-coloured moths, which are better camouflaged, have a selective advantage.	Elephants with small tusks or no tusks are not killed by poachers and so are better able to survive.

»

>>		Antibiotic resistance in bacteria	Peppered moths	Tuskless elephants
	Inheritance	Bacteria that survive pass on the resistance genes to many offspring. The antibiotic-resistant proportion of the population rises. If treated with the same antibiotic again, few die.	Light-coloured moths more often survive to maturity and reproduce. The alleles for light colouration are passed on to the offspring. The frequency of the light-colour allele increases. The frequency of the dark-colour allele declines.	Tuskless elephants survive to reproduce and pass on the tuskless alleles to their offspring. The proportion of tuskless elephants increases in the next generation. (There is evidence that this change is occurring in some populations of African elephants.)

Genetic drift

Genetic drift is the change in allele frequencies over time due to random events. This is particularly evident in small populations. In a small population of prey animals, more animals carrying a particular allele may be eaten than those with the alternative allele. It is not that the particular allele makes them tastier or easier to catch; it is simply chance. Similarly, in a small population, an individual with a rare allele may by chance have more offspring than average, increasing the frequency of that allele.

Bottleneck

Genetic drift can have dramatic effects when a population is quickly reduced in size; for example, after a natural disaster or from over hunting. This results in a **genetic bottleneck**. The allele frequency, and therefore genetic variability, in the population is sharply reduced, meaning that even if the species recovers and grows in number, the new species' members will have less genetic variation than their ancestors. The species becomes more vulnerable to selecting agents in its environment.

Founder effect

When small populations colonise a new area, the individuals in that small colony are unlikely to have the same allele frequencies, due to low genetic diversity, that you would find in the parent population's gene pool. If that new population remains isolated from the rest of the parent population, allele frequencies will vary. This is known as the **founder effect**. The founder effect is important in the frequencies of many inherited diseases; for example, Tasmania has more than three times the incidence of Huntington disease of the USA or the UK.

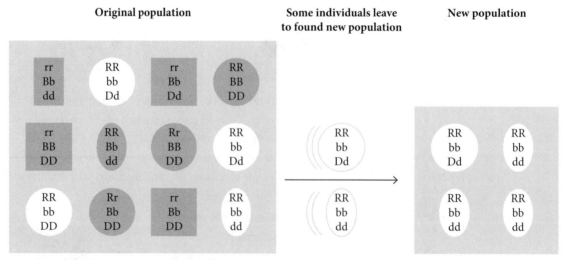

Alleles: R = round, r = square; B = blue, b = white; D = broad, d = narrow

The founder effect occurs when individuals migrate to an isolated area and form a new population with allele frequencies different from the original population.

9780170479431

Gene flow

The movement of individuals out of (emigration) or into (immigration) a population can quickly change allele frequencies. This is known as **gene flow**. Immigration of people from Asian countries to Australia is increasing the frequency of the once locally rare IB allele (blood cell antigen) in the Australian population. Gene flow between populations of plants (which can't migrate!) can occur by the dispersal of seeds, spores or pollen.

Mutations

A **mutation** is any change in the arrangement or amount of DNA in a cell or a virus. Mutations may occur in somatic cells or in germ line cells. Only mutations that occur in germ line cells are inherited through gametes. They can introduce new alleles and increase the variation within a gene pool. Mutations are not always harmful; some can be advantageous, while others may be neutral. They occur spontaneously, but the rate of mutation is increased by exposure to mutagens.

A **mutagen** is any influence that increases the rate of mutation. Physical mutagens include:

- heat
- X-radiation

- gamma radiation
- UV radiation.

Many chemicals are mutagens. These chemicals may modify bases, leading to changes in base pairing. Types of mutations and their effects are summarised below and the next two pages.

Changes in chromosome number

As well as changes to gene structure due to point mutations, page 136, and changes in chromosome structure due to block mutations, page 137, the genetic make-up of an organism can change due to the gain or loss of whole chromosomes or sets of chromosomes. This usually occurs due to failure of the chromosomes to separate correctly during meiosis (called non-disjunction). This results in gametes with more or fewer chromosomes than normal and therefore more or fewer genes for particular traits. These mutations are called chromosomal abnormalities or **chromosomal mutations**. Such mutations would alter the frequency of the alleles for those genes in the gene pool.

There are two types of chromosomal mutations: aneuploidy and polyploidy.

Aneuploidy is a condition where there is one extra or one less chromosome in the genome. Compared to the normal human chromosome number of 46 per body cell ($2n = 23$), Down syndrome individuals have 47 with an extra chromosome 21; Turners syndrome females have 45 with one less X chromosome; and Klinefelters syndrome males have 47 with an extra X chromosome.

Polyploidy is a condition in which individuals have more than the normal 2 sets of chromosomes. This would alter the allele frequencies in the gene pool of a population. There are few surviving cases of polyploidy in animals (salamanders, frogs and leeches are polyploids) and it is a fatal condition in humans so the few that are born die very early. However, polyploidy is common in plants, in which the plants often thrive with such a condition. Examples include macaroni wheat which has 4 sets of chromosomes ($4n$) and bread wheat which has 6 sets of chromosomes ($6n$).

Point mutations

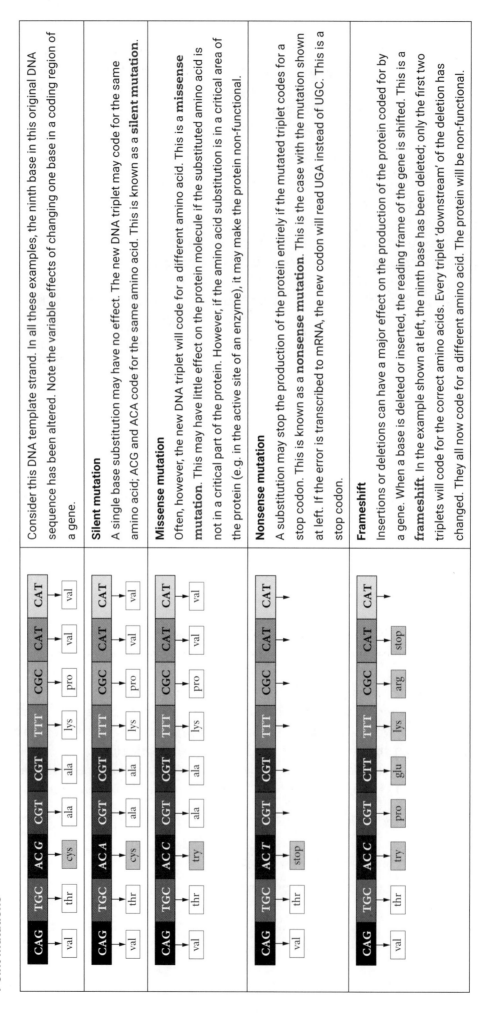

Consider this DNA template strand. In all these examples, the ninth base in this original DNA sequence has been altered. Note the variable effects of changing one base in a coding region of a gene.

Silent mutation

A single base substitution may have no effect. The new DNA triplet may code for the same amino acid; ACG and ACA code for the same amino acid. This is known as a **silent mutation**.

Missense mutation

Often, however, the new DNA triplet will code for a different amino acid. This is a **missense mutation**. This may have little effect on the protein molecule if the substituted amino acid is not in a critical part of the protein. However, if the amino acid substitution is in a critical area of the protein (e.g. in the active site of an enzyme), it may make the protein non-functional.

Nonsense mutation

A substitution may stop the production of the protein entirely if the mutated triplet codes for a stop codon. This is known as a **nonsense mutation**. This is the case with the mutation shown at left. If the error is transcribed to mRNA, the new codon will read UGA instead of UGC. This is a stop codon.

Frameshift

Insertions or deletions can have a major effect on the production of the protein coded for by a gene. When a base is deleted or inserted, the reading frame of the gene is shifted. This is a **frameshift**. In the example shown at left, the ninth base has been deleted; only the first two triplets will code for the correct amino acids. Every triplet 'downstream' of the deletion has changed. They all now code for a different amino acid. The protein will be non-functional.

Block mutations/macromutations

Normal allele	**Duplications**
	Large or small sections of chromosomes may be replicated and so appear more than once in the **genome**. This is known as **duplication**. Examples: globins, trinucleotide repeat expansions.
	The human genome contains several genes that code for different globin molecules. They have evolved from copies of primitive globin genes in our ancestors.
	Increased numbers of trinucleotide repeats in the coding region of a gene result in the addition of many more amino acid molecules in the resulting protein when the gene is transcribed and translated. The function of the protein can be altered.
HD allele	Mutation in the Huntington disease (HD) gene increases the number of CAG repeats in a coding region of the gene.
	Inversions
	Inversions occur when part of a chromosome breaks off and re-joins the chromosome with the DNA 'flipped'. The effects of this depend on where the breaks occur. If they occur in the coding region of a gene or in the promoter region, gene function is lost.
	Deletions
	Deletions occur when part of a chromosome is lost. The chromosome breaks in two places and re-joins, leaving one piece out.
	Large deletions are usually lethal. Smaller deletions can lead to several recognised genetic disorders.
	Translocations
	In a **translocation**, part of one chromosome breaks off and attaches to a different, non-homologous chromosome. The effects vary widely. Sometimes parts of genes are lost in the break. Some translocations have no effect on the carrier, but can be passed on in gametes. This results in the zygote receiving an extra copy of the translocated DNA. A translocation chromosome 14/21 is responsible for some cases of Down syndrome.

6–35 CAG repeats

40–100 CAG repeats

4.1.2 Biological consequences of changing allele frequencies

We know which factors influence changes in allele frequencies, but what are the consequences of this? Imagine a situation in which a population of organisms lives in a stable area over a period of time. This population shows variations in genotype and phenotype. In this imaginary stable world:

- there is **random mating**; each individual is as likely to reproduce as any other and there is no selection of mates based on phenotypes
- all offspring have an equal chance of surviving and reproducing (no selection)
- there is no mutation that alters phenotypes
- there is a large population size
- there is no migration; no members of the population leave the area and no new members arrive.

In such a population there would be *no change* to the frequency of alleles for any gene locus. The alleles present in the population would be passed onto their offspring, who would then have the *same* allele frequencies as the parental generation. Such populations are said to be in genetic **equilibrium**.

Natural populations are often *not* in equilibrium and are said to be in disequilibrium. The frequencies of alleles in populations change over time. Any factor that 'upsets' the equilibrium of a population can cause variations in the predicted allele frequencies in later generations. When allele frequencies and phenotypes change from generation to generation, the population is said to be **evolving**.

Small population sizes are more susceptible than large populations to genetic consequences of decreased allele frequencies within the gene pool. Less variation leads to a decreased ability to respond to selective pressures, which can result in declining population numbers and can increase the risk of **extinction**.

For example, bananas of the same type are genetically identical so, if one plant is susceptible to disease or infection, the entire plantation will also become infected. This was the case in the 1950s when Gros Michel bananas become ravaged by Panama disease. Fortunately, banana companies discovered Cavendish bananas were resistant to Panama disease and they were able to save the industry for another 50 years. However, Panama disease has returned and Cavendish varieties are no longer resistant, leading to concerns that the banana could become extinct.

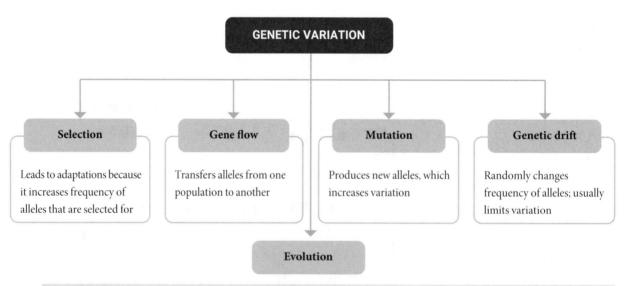

The consequences of factors that influence allele frequencies in a population

4.1.3 Manipulation of gene pools through selective breeding

Selective breeding is a form of artificial selection in which organisms with desirable traits are chosen to breed and others are prevented from breeding. For example, a farmer will want their best milk cow to have many calves and will banish a poor producer. Most breeds of domestic animals have developed their specific phenotypes by selective breeding.

Advantages	Disadvantages
Allows for higher profit	**Loss of variation**
Breeding cows who produce more milk than those typically bred will ensure the farmer will make more money.	Changes in allele frequencies for both the selected traits and the linked traits
	Increased numbers of homozygous organisms; fewer heterozygotes can lead to general lack of physical strength and good health among the population.
Create new varieties of crops	**Cannot control genetic mutations**
Plants of the genus *Brassica* have been bred to produce different foods such as kale, broccoli and cabbage by modifying plant sections through selective breeding.	Random genetic mutations cannot be controlled and may occur, which could decrease effectiveness of the breeding program because most mutations are not beneficial.
No safety concerns	**Desirable traits are not always advantageous**
Selective breeding allows for natural evolution at a faster rate. The DNA sequences are not genetically modified using technology.	It may be that sheep who produce more wool, quicker, grow at much faster rates but their legs may not be strong enough to support the rapid growth.
Eliminate diseases	**Genetic depression**
Identifying and screening for disease allows farmers to not breed from organisms with underlying health concerns.	As organisms become more genetically similar, they are more likely to inbreed, resulting in an increase in negative mutations.
Influence food production	**Evolution of species**
Farmers can produce animals or crops that are better suited to different climates, preserving food supplies.	Once alleles are lost from a gene pool, it is very hard to bring them back. Less biodiversity leads to the greater susceptibility of whole populations of organisms to pathogens and/or changes in environmental conditions.

4.1.4 Genetic change within pathogens and challenges for treatment strategies

Despite non-specific defences, a well-developed immune system and all types of medical interventions, humans still get infectious diseases. Some pathogens have evolved adaptations that make them less likely to be attacked or killed by the body's defences or our medications.

Antibiotic resistance

Antibiotics are chemicals that are used to kill or inhibit the growth of bacteria. The mechanism by which the antibiotic works is dependent on the species of bacteria. The common antibiotic penicillin, for example, inhibits the production of the **peptidoglycan** cross-linking capability of the cell wall, which leads the cell wall to become weakened and can result in **lysis** of the bacteria due to osmotic pressure. However, bacteria have developed defences against antibiotics.

Over many generations, a small proportion of the bacteria will develop antibiotic resistance via genetic mutation. When treated with antibiotics, the resistant bacteria are able to survive and reproduce (via **binary fission**). Antibiotic resistant bacteria may also pass on resistance to antibiotic sensitive bacteria by transferring plasmids that have a gene for antibiotic resistance, resulting in antibiotic resistant bacteria.

The introduction of an antibiotic, a selection pressure, has led to the antibiotic resistance gene to be selected for and become more frequent in bacterial colonies. The reason for the development of these resistant bacteria is over-prescription of antibiotics and misuse of antibiotics; for example, feeling better after a few of days of taking an antibiotic and so not finishing the entire course.

Methicillin-resistant *Staphylococcus aureus* (MRSA), or golden staph, infections are prevalent in settings where methicillin antibiotic use is most common, such as nursing homes and hospitals. Doctors now prescribe alternative antibiotics to treat infections caused by MRSA, such as vancomycin.

Enterococci are a different genus of bacteria, some of which have developed resistance to the antibiotic vancomycin. These resistant bacteria are called VRE, or vancomycin resistant *enterococci*. They are a major cause of infection in hospitalised patients, particularly in intensive care units, as they have the ability to colonise the gastrointestinal tract of humans and survive in the environment on surfaces.

Treatment of patients who acquire VRE requires a prolonged stay in hospital and usually the administering of different (i.e. not vancomycin) intravenous antibiotics. VRE are associated with high mortality rates and high health and economic costs. Therefore, preventing the spread of VRE is a priority and strategies include better hand hygiene, active screening and superior cleaning of surroundings.

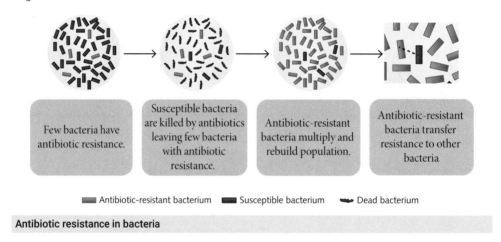

| Few bacteria have antibiotic resistance. | Susceptible bacteria are killed by antibiotics leaving few bacteria with antibiotic resistance. | Antibiotic-resistant bacteria multiply and rebuild population. | Antibiotic-resistant bacteria transfer resistance to other bacteria |

▬ Antibiotic-resistant bacterium   ▬ Susceptible bacterium   ∿ Dead bacterium

Antibiotic resistance in bacteria

Antigenic variation

Influenza is caused by three main types of virus, but the vast majority of cases (about 95%) are caused by RNA viruses known as type A influenza virus. Like all RNA viruses, the flu virus has a core of RNA. It is an enveloped virus, meaning it has an outer membrane that is derived from its host.

This membrane is studded with two viral proteins: **haemagglutinin (H)** and **neuraminidase (N)**. The H and N proteins vary between strains of influenza virus. This is what the numbers mean when specifying strains of flu. There are 18 haemagglutinin subtypes and 11 neuraminidase subtypes (H1–H18 and N1–N11, respectively). The 2009 swine flu was an H1N1 virus, meaning H type 1 and N type 1. The 2015 seasonal flu was caused an H3N2 virus.

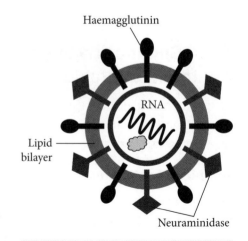

The structure of a type A influenza virus

Some populations of pathogens are very prone to mutation and hence change their surface antigens as they grow and reproduce inside a host. The body has no immunity to the new antigen. This is particularly true of influenza viruses, where novel versions of the surface antigens occur every year. New combinations of antigens can occur when more than one virus is present in an organism at the same time. For example, if a pig is infected with both the bird flu and human flu viruses, the viral genes can be rearranged to make new combinations in the new viral particles.

Antigenic drift vs antigenic shift

Influenza virus is able to change via **antigenic drift**. These are small mutations in the genes of the virus that result in changes in the antigens HA (haemagglutinin) and NA (neuraminidase) on the surface of the virus. Influenza viruses that are closely related to one another are usually antigenically similar enough that the antibodies your immune system creates against one influenza virus can respond to a similar influenza virus. This is known as **cross-protection**.

However, with antigenic drift happening continually over time as the virus replicates, the small changes can produce viruses that are still closely related to one another but are antigenically different enough that the antibodies your immune system produced in response to the original virus will not be able to recognise the new one. Antigenic drift is the main reason people become susceptible to flu infections even after they have been infected. It is also why the flu vaccine must be reviewed and updated each year to keep up with the evolving influenza virus.

Antigenic shift is another change in the HA and NA antigens on the surface of the influenza A virus. Antigenic shift is a much more sudden and dramatic change in the antigen combinations compared to antigenic drift. It can occur when influenza viruses from an animal population are able to infect humans. The process of diseases from animals infecting humans is known as zoonosis. Animal-origin viruses can contain HA or HA/NA combinations that are antigenically different from those found in humans. This means most people do not have immunity to the new virus. Examples include the viruses that cause swine flu (H1N1) and COVID-19 (SARS-CoV-2).

Antigenic shift is seen less often than antigenic drift but, when it does occur, it can result in a pandemic. As we have all witnessed with COVID-19, the challenges resulting from antigenic shift include health crises, with a race to protect people through careful hygiene, screening, quarantine and isolating measures; and economic crises caused by job losses and job cuts. Further, there is a race to develop a safe and effective vaccine before the virus is able to shift and mutate again.

4.2   Changes in species over time

4.2.1 Changes in species over geological time

A **fossil** is the **remains** of an organism or direct evidence of its presence (e.g. footprints, casts or moulds) that has been preserved. Fossils may be preserved in rock, ice, amber, tar, peat or volcanic ash. The most common fossils are of the hard parts of organisms, such as bones, teeth and shells. In order for an organism to be fossilised, the following conditions must be met.

> **Hint**
> Remember the acronym RUDD (Rapid burial, Undisturbed, Decomposition, Downward pressure) for the necessary conditions for fossilisation.

Rapid burial	Animals may be trapped in a bog, rapidly covered in sediment, frozen in ice, trapped in tree sap that hardens to amber, fall into a peat bog or tar pit, or be caught in volcanic ash.
The organism lies undisturbed	Fossilisation often involves the replacement of animal or plant tissue with hard minerals deposited from the surrounding rocks and sediments. This is a slow process that relies on the remains being undisturbed for many hundreds or thousands of years.

Decomposition is prevented	Conditions must be such that bacteria that normally decompose dead tissue cannot survive. These conditions include extreme cold, dehydration, high acidity or the absence of oxygen. Soft tissue decomposes first; being moist, it can be lost by simple chemical breakdown, explaining why fossils of internal organs are rarely found.
Downward pressure	A high-pressure environment promotes the mineralisation of remains. This is where organic matter, or bone, is replaced with minerals from the environment, actually turning the body parts to hard rock.

Fossils have provided evidence of the evolution of modern species from ancestral forms. The evolution of modern horses from a primitive four-toed mammal has been shown from fossil evidence. Another famous fossil is that of *Archaeopteryx*, an animal that shows features of both reptiles and birds. Fossil evidence has also shown us that modern whales descended from a terrestrial, hoofed mammal.

For much of the history of life on Earth, life forms were soft-bodied invertebrates, unlikely to leave much fossil evidence. The table below shows a brief history of life on Earth in the Phanerozoic eon, a period for which there is a great deal of fossil evidence. This eon began with the Cambrian explosion – a rapid increase in the size and diversity of life on Earth. The Phanerozoic eon is divided into three eras: the Palaeozoic (ancient life), Mesozoic (middle life) and Cenozoic (modern life). Some important evolutionary steps are shown in the table. You do not need to know the names of all of the periods and epochs, but you should be familiar with the overall history of vertebrate evolution.

Evolutionary timeline

Time (years ago)	Evolutionary step
4.6 billion	Earth is formed
3.7 billion	First prokaryotic cells
3.4 billion	First photosynthetic cyanobacteria
1.8 billion	First eukaryotic cells
1 billion	First simple multicellular organisms
600 million	First multicellular animals
500 million	First chordates, simple fish and proto-amphibians

Plant evolution

Eon	Era	Period	Events	Reproduction
PHANEROZOIC	Cenozoic	Quaternary	Rapid evolution of the grasses, a group of angiosperms.	Angiosperms produce seeds with a protective coating, enabling them to reproduce in most environments.
		Tertiary	Rapid diversification and spread of angiosperms to most environments, corresponding to an explosion in insect variety.	
	Mesozoic	Cretaceous	The first angiosperms (flowering plants) appear. Forests of gymnosperms dominate Earth: conifers (pines), cycads (palms) and gingkoes.	Gymnosperms reproduce by seeds ('gymnosperm' means naked seed), protected in cones and similar structures.
		Jurassic		
		Triassic		

PHANEROZOIC	Palaeozoic	Permian	The first seed plants appear; ancestors of gymnosperms.	Distribution of fern forests is limited by the need for a moist environment for spores.
		Carboniferous	Tree ferns form large forests.	
		Devonian	First vascular plants appear; ancestors of ferns and tree ferns.	Plants still rely on a watery environment for the dispersal of spores for reproduction.
		Silurian	First land plants appear. They are non-vascular.	
		Ordovician	The age of algae	Plants reproduce in water by sexual and asexual means.
		Cambrian		

Dreamstime.com/Dario Lo Presti

Animal evolution

Eon	Era	Period	Epoch	Events
PHANEROZOIC	Cenozoic	Quaternary	Holocene	Recent times
			Pleistocene	Ice Age Large mammals peak and then become extinct
		Tertiary	Pliocene	Earliest humans
			Miocene	Diversity of grazing mammals and songbirds
			Oligocene	First apes
			Eocene	All modern orders of mammals appear Radiation of bird groups
			Palaeocene	Explosion of primitive mammals

››

		Cretaceous	Dinosaurs peak and then become extinct	
PHANEROZOIC	Mesozoic	Jurassic	Dinosaurs dominate Earth First birds	
		Triassic	First mammals First dinosaurs	
	Palaeozoic	Permian	Landmasses joined in a supercontinent, Pangea	
		Carboniferous	First reptiles	
		Devonian	First amphibians Many trilobites	
		Silurian	First terrestrial arthropods Coral reefs are common	
		Ordovician	First fish First land plants	
		Cambrian	The age of marine invertebrates First chordates appear	

The law of faunal succession

Sedimentary rock type is where most fossils are found. The surface of Earth is arranged intro sedimentary layers (stratification) where the oldest stratum, or layer, is at the bottom and new layers form on top. Different regions will not always have the same sedimentary layers due to erosion, flooding or other environmental impacts.

The law of faunal succession is based on the *observations* of the **fossil record**, where the sedimentary rock layers, or strata, contain fossils that supersede each other as you move up the layers. The succession of the fossils is specific and reliable. For example, *Homo habilis* would never be found in the same stratum, or layer, as a tyrannosaurus because they lived in different geological periods. This enables strata to be identified and dated by the fossils that are found within it.

The fossil record also shows that features of organisms change over time, and that different organisms do not occur randomly but are the result of evolution.

Transitional and index fossils

While the fossil record provides hints about evolutionary relationships, there are many gaps in the record. In fact, it is estimated that the number of species known about through fossils is less than 1% of all the species that have lived. This is due to several reasons, including the rare environmental conditions required for an organism to become fossilised. Hard body parts are more easily preserved, so there are more fossilised vertebrates than invertebrates, and very few fossils have been found.

Transitional fossils are the remains of species that exhibit traits common to their ancestral group and their predicted descendants. The *Archaeopteryx* is an example of a transitional fossil because it shows the links of evolution between dinosaurs and birds; the fossil shows that it possessed jaws, claws and feathered wings. As new fossils are discovered, these patterns of evolution are continually challenged and re-evaluated.

Index fossils, or guide fossils, are indicative organisms, as they are short-lived species and can be found in specific rock strata spread over a wide geographic area. They can be used to determine the relative age of the rock layers. For example, ammonite fossils indicate an age range of about 440 million years ago to 360 million years ago.

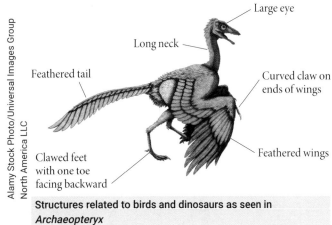

Alamy Stock Photo/Universal Images Group North America LLC

Large eye

Long neck

Feathered tail

Curved claw on ends of wings

Clawed feet with one toe facing backward

Feathered wings

Structures related to birds and dinosaurs as seen in *Archaeopteryx*

Fossil dating

The age of a fossil is determined in one of two ways: comparative age (the relative age of fossils can be roughly determined by the sediment layers in which they occur) and absolute age. The absolute age of fossils can be determined using radiometric dating.

Absolute dating

Radiometric dating is based on the fact that many elements in Earth's atmosphere and rocks exist as different isotopes (atoms have different numbers of neutrons). Some of these isotopes are unstable and emit radiation (they are radioactive). These radioactive isotopes decay over time, giving off neutrons and/or protons, and become more stable atoms. The radioactive isotopes decay at a known, constant rate. The time in which half of a sample of a radioactive isotope will decay to a more stable form is known as its **half-life**. One isotope commonly used in dating is carbon-14, which has a half-life of 5730 years.

When an animal dies, the proportion of carbon as carbon-14 in its cells is the same as the proportion in the atmosphere. The shaded circles represent atoms of carbon-14 among atoms of the more common carbon-12 at the time of an animal's death.	
Analysis of a fossil thousands of years later shows that the proportion of carbon-14 is now only one-eighth of what it was. The rest has decayed. This information can be used to determine the age of the fossil.	
As only one-eighth remains, the fossil has existed for three half-lives. $1 \rightarrow \frac{1}{2} \rightarrow \frac{1}{4} \rightarrow \frac{1}{8}$ As one half-life is 5730 years, the fossil is 3 × 5568 years old. The fossil is approximately 17 190 years old.	

Carbon-14 has a relatively short half-life, so it can only be used to date relatively recent fossils, up to about 50 000 years old. Other radioactive isotopes have longer half-lives and can be used to date older fossils and sediments.

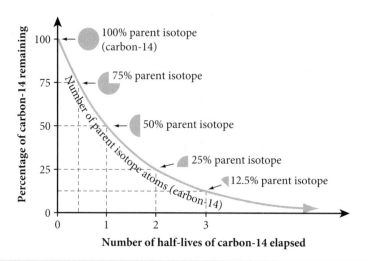

Graph showing the radioactive decay over time

Isotope	Decays to:	Half-life	Age of deposits it can determine
Carbon-14	Nitrogen-14	5730 years	100–50 000 years
Uranium-235	Lead-207	704 million years	More than 500 000 years
Potassium-40	Argon-40	1.3 billion years	More than 500 000 years
Rubidium-87	Strontium-87	48.8 billion years	More than 100 million years

Electron spin resonance (ESR) is useful for dating more recent organic samples that radiometric dating cannot, as the sample may be too old or too young. ESR is commonly used to date tooth enamel, coral and barnacles that are between 5000 and 2.4 million years old. It is an absolute dating method that depends on radiation from the soil. The radiation causes electrons in minerals to move to a higher energy state, and the number of high energy electrons in the sample can be used to determine when the sample was buried. By measuring how many high energy electrons there are in the sample, we can calculate how long it took for them to accumulate, and hence deduce when the sample was buried.

Relative dating

Relative dating is used to sequence geological events using the rocks they leave behind. This method of reading the layers of rock, or strata, is known as stratigraphy. Relative dating provides a rough estimate of the dates and not the exact numerical age.

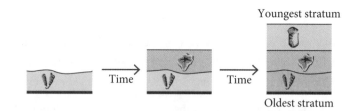

The bottom layers of the rock are the oldest.

4.2.2 Evidence of speciation

So far, we have looked at variation within species and microevolution. It is not a huge step to imagine the forces that drive microevolution, especially natural selection, over long time periods, causing such significant changes in populations that they are no longer recognisable as the same species. Over many generations, members of the population develop adaptations to different selection pressures and hence evolve differently. This is described as **divergent evolution**. Over even longer periods of time, ancestral populations can evolve into different genera, families, orders, classes, phyla and even kingdoms. The evolution of new species and other taxa is known as macroevolution, and the formation of two or more species from one ancestral population is known as **speciation**. **Convergent evolution** is different to divergent evolution. Convergent evolution is when unrelated organisms faced with similar selection pressures evolve similar adaptations, for example, bats and dolphins are not closely related but both have evolved the ability to echolocate.

Speciation leads to new species

A species is a population of organisms that are phenotypically similar *and* are capable of interbreeding to produce viable, fertile offspring. Different populations of organisms are considered different species if they do not successfully interbreed, even if they look very much alike. There is no gene flow between species, so different species are said to be **reproductively isolated** from each other. There are some problems with this definition. What if two populations never encounter each other in the wild and never mate, but they will mate and produce viable offspring in a laboratory? Where do we draw the species line in populations of organisms that reproduce asexually? Are two populations different species if they exist in different times? Despite the flaws, this definition is the best we have to date.

There are a number of steps along the way to the evolution of new species from existing species.

Allopatric speciation in the Galapagos finches

The term 'allopatric' comes from the English word for 'other' (*allo*) and the Greek word for 'fatherland' (*patria*). This type of speciation is the evolution of one or more new species from a single species as a result of the original species becoming geographically isolated by a barrier that prevents or severely limits gene flow. The species undergoes genetic divergence due to a difference in selection pressures on either side of the barrier. A barrier could be a river, mountain range, ocean or changing landscape.

The table below outlines the allopatric speciation that occurred in the Galapagos finches.

An ancestral species of finch colonises an island in the Galapagos. They live across a wide range, but gene flow occurs through the population.	
Some finches move to other islands and are now separated from the other finches by a **geographical barrier**, the ocean. Gene flow between the populations stops.	
Selection pressures such as food sources on the islands are different, leading to selection for different phenotypes. As mutation is a random event, you would expect different mutations to arise in the two populations. The two populations evolve different beaks that are adapted to their environments.	
If the barrier remains and the subspecies evolve in different directions, they could become reproductively isolated; they are now different species.	
Evolution of a new species can be more rapid if only a small population is isolated by the barrier. The effects of natural selection, founder effect and genetic drift can lead to rapid speciation.	

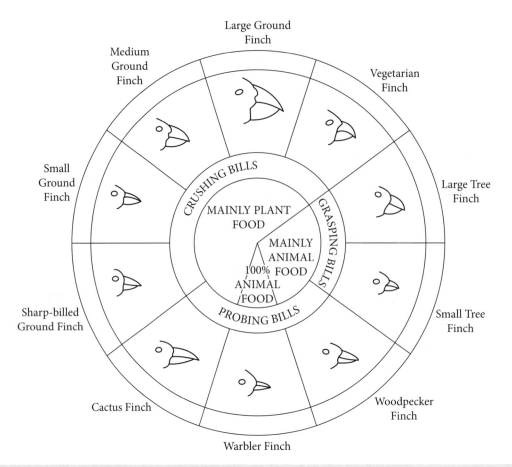

Beak types in Galapagos finches based on the food source available

From the picture above, you can see that the ancestral species has diversified into many new species as the change in the environment has impacted the resources available and creates new environmental niches. This process, which leads to many new species, is known to evolutionary biologists as **adaptive radiation**.

Sympatric speciation in *Howea* palms on Lord Howe Island

'Sympatric' speciation, derived from the English word for 'together' (*sym*) and the Greek word for 'fatherland' (*patria*), is the genetic divergence of a species that lives in the same geographical region but evolves separately until its members can no longer interbreed. In the example of *Howea* palms on Lord Howe Island, there is evidence to support sympatric speciation in a case study of two species of palm (*Howea forsteriana* and *Howea belmoreana*).

A DNA-based phylogenetic tree for the subfamily of palms *Aracoideae* supports the monophyly (shared common ancestor) of the two palm species and shows that they diverged from each other well after the island was formed 6.9 million years ago.

The following table outlines the sympatric speciation that occurred in the palms of Lord Howe Island.

Origin species of palms growing on Lord Howe Island	
Soil types and pHs provide microhabitat differences that provide selection pressures allowing members of the same species to thrive in different ways.	
Plants that grow best in basic calcareous soil flower earlier than those growing in volcanic, acidic soil.	
The species becomes reproductively isolated and diverges into two distinct species sharing the same geographical area.	

Reproductive isolation

Populations may become reproductively isolated due to **prezygotic** barriers (factors that prevent reproduction) or **postzygotic** barriers (factors that prevent the survival of hybrid offspring).

Prezygotic barriers include:

- differences in reproductive seasons (sorry, I'm not fertile at the moment!)
- differences in the structure of genitalia (sorry, it won't fit!)
- incompatible behaviours (sorry, your mating call does nothing for me!)
- gametic isolation (egg and sperm have chemical differences that prevent fertilisation).

Postzygotic barriers include:

- hybrid inviability: inability of any hybrids to survive
- hybrid sterility: hybrids cannot produce viable gamete
- hybrid breakdown: offspring of hybrids are unable to reproduce.

4.3 Determining the relatedness of species

4.3.1 Evidence of relatedness between species

A comparison of the anatomy of different species provides evidence of relatedness between species. The presence of **homologous structures** and similar **embryonic development** in many organisms indicates descent from a **common ancestor**. (In this context 'common' means 'shared', not 'ordinary'.) **Analogous structures** and **vestigial structures** underline the importance of natural selection in determining the development of species.

	Evolutionary origin	Examples
Homologous structures	Structures that have a common evolutionary origin and similar underlying anatomy, but have evolved in different ways in different groups due to different selection pressures	Forelimbs of different mammals

	Evolutionary origin	Examples
Developmental biology	In the early stages of development, the embryos of diverse groups are very similar. Even humans have gill slits and tails. The explanation is that the genes that direct the development of these structures have been inherited from a common ancestor. Many embryonic structures are lost in later development due to the action of other genes.	Early embryonic stages of (from left to right): fish, turtle, rabbit and human
Analogous structures	Structures that have similar functions but have evolved independently. The same selection pressures have caused the same feature to be selected for in diverse species.	Body shape of sharks (fish) and dolphins (mammals)
Vestigial structures	Structures that are reduced in size and role due to lack of selection pressure for their maintenance. The structure remains part of the organism's anatomy despite having no useful purpose, because there is no selection pressure for its complete removal.	The pelvis and miniature leg bones of whales and snakes

Molecular homology

Molecular homology is the comparison between nucleotide sequences or amino acid sequences of polypeptides as evidence for common evolutionary origin. Some of the biochemical techniques used to determine the evolutionary relationships between species are discussed below.

DNA sequencing

The sequence of nucleotides in related genes in different species can be determined. As mutations in a gene tend to accumulate over time, the number of differences in the base sequences in the same gene in different species can give an indication of the relatedness of the species. The more mutations there are, the longer the species have been separated.

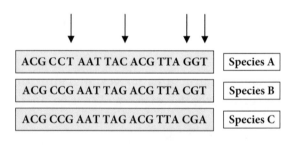

Consider the three hypothetical base sequences. Species A and B differ in three bases, species B and C differ in only one base. Species B and C are more closely related to each other than either is to species A.

mtDNA sequencing

Mitochondria contain DNA that is separate from the DNA in the nucleus. Mitochondria, and hence their DNA, are passed from mother to egg to child, so mtDNA is inherited solely from the maternal line. mtDNA mutates at a faster rate than nuclear DNA, so comparisons can be made between closely related species or different populations of the same species.

mtDNA and nuclear DNA are different in structure and hence have different uses in scientific application: mtDNA is closed and circular, whereas nuclear DNA is linear and open-ended.

Mitochondrial DNA acts as a molecular clock; the more mutations a population has accumulated over time, the longer its time of evolutionary change. Nuclear DNA has a lower mutation rate, so it provides less information in phylogenetic applications. However, scientists use nuclear DNA in forensics because it contains the genetic blueprint: it is inherited from both the mother and father and is unique to each individual.

Protein analysis

The sequence of amino acids in protein molecules can be determined. Where there are differences between species in the amino acid sequence of a protein, we can infer that these differences are due to mutations in DNA. The number of different amino acids is an indicator of the number of mutations and hence the degree of separation of species.

4.3.2 The use and interpretation of phylogenetic trees

The study of evolutionary relationships and their development is known as **phylogeny**. In order to represent the relationships between groups or species of organisms, we can use and interpret phylogenetic trees.

Phylogenetic trees are diagrams that represent hypothetical evolutionary relationships, based on how a group or several species evolved from a common ancestor. The hypothesis can be based on morphological traits such as an organism's form or structure, or on molecular data such as DNA or amino acid sequencing.

	Advantages	Disadvantages
Morphological data	Ability to hypothesise and understand the morphology and ecology of extinct, ancestral species Can estimate time of divergence of lineages using fossil data	Morphological traits are more likely to converge The fossil record is incomplete so there is less data Patterns of inheritance may not be clear and can be subjective
Molecular data	Lots of data Less likely for the traits to show convergence because, even if phenotypes are similar, the molecular data could show differences; this is because there are multiple sequences that could produce the same phenotype due to the redundancy of the genetic code Can use non-coding regions of DNA, which are not under selection and will mutate at a faster rate The number of mutations can be used to measure time, as we can calculate the rate of mutation No need to rely on fossil data	No information can be obtained on the morphology and ecology of extinct species

Interpreting phylogenetic trees

This diagram shows a phylogenetic tree where the common ancestor, shown as the root, diverged into three separate species. The incoming line on the root indicates that the species originates from a larger clade. A clade includes a node, with all of its connected branches representing a common ancestor that descended into different species. A clade is also sometimes referred to as a monophyletic group because it can be separated from the tree with a single cut. Here, the root species descended into two groups. The first group then split again into two other

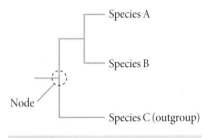

Interpreting phylogenetic trees

species. The node is the common ancestor of the descendent species. The outgroup is the most distantly related species, which acts as a reference.

Phylogenetic trees can be represented in many formats, but the relationships between groups will not change. For example, once you identify the root, and rotate the tree using the root as a pivot, you can see that the relationships between species of interest remain the same.

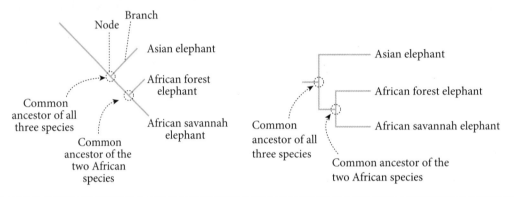

Phylogenetic trees may be represented differently. Clades include a common ancestor and all its descendants.

4.4   Human change over time

4.4.1 Shared characteristics

Humans are **primates**, members of one of the oldest orders of mammals. **Mammals** are characterised by fur, mammary glands to feed their young, and three bones inside the inner ear. Primates evolved from mammals with features that are advantageous to arboreal (tree) life. Primates separate out into two distinct groups: monkeys and apes. The great apes, including humans, are **hominoids**, characterised by large brains and lack of a tail. Modern humans are the only surviving species of **hominin** (human and human-like), but there have been others in the past, some ancestral to us and other extinct branches on our family tree. Hominins have evolved structures that enable them to stand erect and have bipedal locomotion such as the position of the foramen magnum in the skull, curvature of the spinal column and the shape of the pelvis.

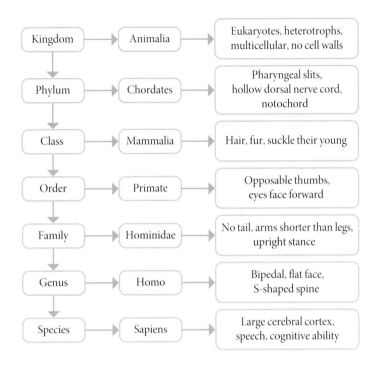

4.4.2 Trends in evolution

Hominins have inhabited parts of Earth for about 4.4 million years. A number of features separated the early hominins from their ape relatives. *Australopithecus* species, found in Africa, have been dated to around 4 million years go. Their fossils show they walked upright but also climbed trees. They had both human and ape features such as a projecting chin, flat nose, small cranium, long arms and long fingers, which were used to climb trees. It was previously believed that a large brain separated the early hominins from the other apes, but it now seems that bipedalism, the ability to walk on two legs, was our first step (no pun intended). Most of the features that separate hominins for apes are related to bipedalism, which in turn is related to a change in habitat from an arboreal way of life to life on the forest floor or open grasslands.

The major trends in evolution are outlined below.

	Australopithecus	*Homo habilis*	*Homo erectus*	*Homo neanderthalensis*	*Homo sapiens*
Brain size	A mutation in the protein of jaw muscles gave rise to smaller muscles around the jaw and skull, allowing the brain more room to develop and grow. The increase in the cranial capacity, which is a measure of the volume of the interior of the skull, is an indicator of brain size. A larger brain indicates higher intelligence and cognitive capacity and ability to create a complex language, belief systems and culture.				

A. afarensis

H. habilis

H. sapiens

>>

››	**Skull**	The face changes to a flatter structure, with a less prominent brow ridge. The position of the foramen magnum, the hole in the skull for the spinal cord, moves from the back of the skull to the centre. Foramen magnum
	Teeth	Hominins have a parabolic shape, and there is no gap between the front and back teeth. There is a reduction in the size of the front teeth, especially the canines. Gorilla *Homo erectus* *Homo sapiens*
	Walking	There is a shift to a more upright, bipedal movement. In *Homo sapiens*, the angle of the femur is inward, which helps to ensure the feet and knee are underneath the body's centre of gravity to aid balance when standing on one leg. Gorilla *Homo erectus* *Homo sapiens*
	Limb structure	There was a shift to longer legs and shorter arms; alignment of the big toe with the other toes. Compared to other primates the human wrist is flexible and the human hand is straight and broad. It has relatively short, straight fingers with a long, strong thumb. The thumb can extend to each of the fingertips providing it with a precision grip. This enables sewing, writing and manipulation of objects such as tools with exquisite dexterity.

Spine	The curvature of the spine changed from a C-shape to a more S-shape. This allows the organism to bear weight better when standing on two feet.
	Gorilla Human
Pelvis	A shorter and broader, bowl-shaped pelvis evolved to support organs in an upright position. The dotted lines show the direction weight is transmitted.
	Chimpanzee Human
Feet	Feet bones have become curved, resulting in arched feet compared to the flat foot of apes. Toe bones are short and straight, and the big toe is straight and no longer opposable, but this helps provide more thrust when walking.
	Human Gorilla

The field of human evolution is one of the most controversial areas in Biology, not least because it involves *us*! Here are some reasons for controversy.

- Most of the information about the early hominins comes from fossils that are incomplete.
- There is often disagreement about whether a species is an ancestor of modern humans or an extinct branch, or dead end, of the human line.
- There is disagreement about which features are the most significant in determining the evolutionary line to modern humans. Should a species be placed on the modern human line if the skull and jaw appear more modern, but the limbs are more ape-like? Or is a species with more human-like limb proportions but a smaller brain ancestral to modern humans?

4.4.3 Evidence for classification

Fossil evidence suggests that modern humans, *Homo sapiens*, have existed for about 180 000 years. The oldest fossils have been found in Africa. Fossils from the Middle East have been dated at approximately 100 000 years, those from China and Australia at 60 000 years and Europe at 35 000 years.

There are two hypotheses concerning the origins of modern humans:

- the **'out of Africa' hypothesis**, which states that modern humans originated in Africa and migrated to all other parts of the world, and divided into several different species

- the **multi-regional hypothesis**, which states that modern humans evolved from ancestral populations of other *Homo* species in many parts of the world and had recurrent gene flow.

As there is more than one hypothesis, different species names and/or different family trees can be produced from the information. The following table is a summary of information about the widely recognised species of hominins.

	Time (yr BP)	Features	Location
Orrorin tugenensis	6 to 5 M	Probably bipedal, but with arms adapted for tree climbing. *Homo*-like teeth and jaws. Not universally accepted as a hominin.	Africa
Ardipithecus ramidus	4.4 M	Little is known as few fossils have been examined. Bipedal, about 1.2 metres tall.	Africa
Australopithecus anamensis	4.2 to 3.9 M	Bipedal, ape-like jaw but teeth like later hominins. Limb proportions more like apes than later humans.	Africa
Australopithecus afarensis	4 to 3 M	Bipedal, small brain, pronounced supraorbital ridges, large jaw and canines. Limb proportions more like apes than later humans.	Africa
Australopithecus garhi	2.5 M	Bipedal, skeleton like later *Homo*, skull like earlier *Australopithecus*. Probably a tool user and meat eater.	Africa
Australopithecus africanus	3 to 2.1 M	Bipedal, hands and teeth distinctly human-like, probably omnivorous. A branch on the human tree, probably not ancestral to Homo, but with limb proportions more human than ape-like.	Africa
Paranthropus species	2 to 1.2 M	Larger jaws and supraorbital ridges than *Australopithecus*; bony ridge on the skull. These species (at least three, possibly more) are all extinct and not direct ancestors of modern humans.	Africa
Homo habilis	2.3 to 1.6 M	Larger brain and smaller teeth than *Australopithecus*. Made and used tools. Limb proportions more like apes than modern humans.	Africa
Homo rudolfensis	2.2 to 1.8 M	Larger brain than *Australopithecus*. Very large molars. Some paleoanthropologists group these fossils in *H. habilis*, others in a different genus, *Kenyanthropus*.	Africa
Homo ergaster	2 to 1 M	Larger brain, the first hominin to leave Africa; hunters, used fire. Limb proportions like modern humans.	Africa, Eurasia
Homo erectus	1.8 M to 300 000	Larger brain, advanced tool making, widespread in Asia, with some populations surviving until very recently. Although the skull and face differ from modern humans, the rest of the skeleton is very similar.	Asia

	Time (yr BP)	Features	Location
Homo heidelbergensis	800 000 to 100 000	Previously called archaic *Homo sapiens*. Larger brain, more human-like but still with sloping forehead, no chin and heavy brow ridges.	Africa, China and Europe
Homo neanderthalensis	400 000 to 36 000	Largest brain. Sloping forehead, heavy brow ridges, chin, heavily built. Advanced social system, burial of dead. Studies of mtDNA support the hypothesis that they are a different species from modern humans and an evolutionary dead end.	Europe, Middle East, Russia
Homo denisovans	300 000 to 40 000	This species has been described only from DNA from two tiny fossils, and hence is said to be putative (is generally thought to have existed). The DNA shows that Neanderthals and Denisovans are very closely related, and that their common ancestor split off from the ancestors of modern humans about 500 000 years ago. Denisovans split from Neanderthals about 300 000 years ago.	Asia, Middle East? South East Asia?
Homo sapiens 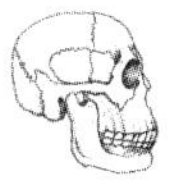	180 000 to ?	Large brain, thin skull, forehead, chin, flat face, small teeth – us.	Everywhere!

Controversies in human evolution

Given that our knowledge of human evolution is based on fossil analysis, different interpretations of evolutionary links are possible. The following diagram summarises human evolution and includes some explanation of the continuing areas of controversy. Note that some species are considered ancestral to modern humans and others are clearly evolutionary dead ends.

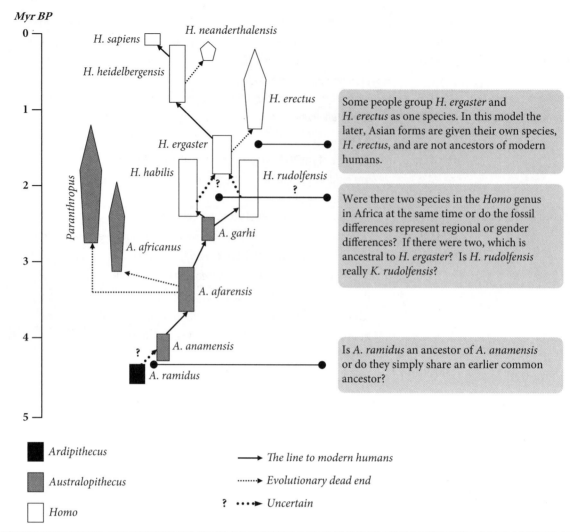

Human evolutionary links based on fossil evidence

There is often disagreement about whether a new fossil belongs in a new species or represents regional variation in the same species. Some people tend to be 'lumpers', grouping many different fossils in the same species and explaining differences in terms of regional variation, sexual dimorphism (differences between males and females) or individual variation. 'Lumpers' may question whether in a couple of million years' time, our distant descendants unearthed the remains of our biggest AFL ruckman and our most petite gymnast, would they classify them in the same species?

Others tend to be 'splitters', using differences in the features of fossils to assign them to different species or genera. They argue that the human evolutionary tree is more like a bush, with many branches leading to evolutionary dead ends.

Is there another whole group of hominins with a strong claim to be the true ancestors of modern humans? Some paleoanthropologists recognise another genus, *Kenyanthropus*, in the evolution of hominins. A 3.5-million-year-old fossil, found in Kenya in 2000, has ignited the controversy. Some researchers (the 'splitters') claim it is significantly different from specimens of *A. afarensis* of a similar age, especially in that it appears to have a flatter face, and assign it to a new genus, *Kenyanthropus*. The 'lumpers' say it is just an example of regional variation and that the differences are magnified by the poor condition of the fossil. So we have one fossil with two names: *Kenyanthropus platyops* and *Australopithecus afarensis*.

Another fossil, found in 1972 and in reasonably good condition, was determined to be about 1.8 million years old. It was assigned to be *Homo habilis* by the 'lumpers', but given the new species name *Homo rudolfensis* by the 'splitters'. It too had a flatter face than other fossils of the same age. In the light of the 2000 discovery of *Kenyanthropus platyops/Australopithecus afarensis*, some researchers now consider this 1.8-million-year-old fossil to be a descendant of *Kenyanthropus platyops* and name it *Kenyanthropus rudolfensis*. Yet another paleoanthropologist thinks it is an Australopithecine and named it *Australopithecus rudolfensis*. One fossil, an age agreed by all, and four species names!

New family members

Not so long ago it was believed that fossil forms of hominins represented a straight line of evolution from early forms to later forms, with earlier species sliding to extinction as new versions evolved. Since the advent of accurate dating, the proliferation of fossil finds and the growth of DNA analysis, this simple line model has been overturned.

The human evolutionary tree is more like a bush, with many branches bearing different species that lived at the same time and sometimes in overlapping regions. A time traveller to Africa 2.5 million years ago could encounter up to five different species of hominins, some ancestral to modern humans, some completely extinct, and perhaps others who interbred with modern humans' ancestors before themselves becoming extinct. Pop into the Middle East about 200 000 years ago and you'd find Neanderthals, Denisovans and the ancestors of modern humans, and in Asia you'd find Denisovans not far from remnant populations of *Homo erectus*, and later mingling with modern humans. As new discoveries of fossils and DNA analysis arise, the classification of humans is continually challenged and refined to best fit the evidence.

The table below details some of the more recent discoveries of agreed species on the human evolutionary 'bush'.

Homo floresiensis	In October 2004, an exciting find that sent human evolution into a spin was announced. Fossils of a new species, named *Homo floresiensis*, were discovered on the Indonesian island of Flores. The fossils were thought to be only 18 000 years old and in many features resembled *Homo erectus*, but were very small. An adult female was only about one metre tall and had a cranial capacity of only 417 cm^3. Despite its small brain, it was a toolmaker and user, and it killed and butchered animals. Unwilling to accept that there was another hominin species around so recently, some people have interpreted these remains as those of modern humans with a disease or condition that has affected brain size and stature. It now seems clear, however, that this species was related to *Homo erectus*, and despite its small brain was intelligent and had a well-developed culture. In 2016, further study cast doubt on the original dating and it now seems that the species may have died out about 50 000 years ago.
Homo denisovans	In 2008, a group of Russian researchers explored the Denisova caves in Siberia. They found a finger bone of a hominin dated at about 41 000 years old. Years before, a hominin tooth had been found at the same location. The finger bone was well enough preserved for the extraction and sequencing of its DNA. The DNA revealed that the fossil represented a new species, different from both earlier species and modern humans. It was determined that this new species, *Homo denisovans*: • diverged from the modern human line about 500 000 years ago • was related to Neanderthals, diverging from them about 300 000 years ago • probably resembled Neanderthals; the one fossil finger was robust, and the two species share a lot of DNA.
Homo naledi	In 2013, a huge collection of the remains of several hominins was discovered deep in a cave in South Africa. The fossils contained jaws, teeth, limb bones, skulls and other fragments. The fossils presented a few surprises: the skull was modern in shape, but the brain case was very small; the shoulders and arms appeared adapted for tree climbing; the hands and wrists were human-like, but with curved fingers; the hips were primitive but the legs and feet were very human-like, indicating that they walked upright. As yet, it has not been possible to date the fossils. Stay tuned for how this species may shake the bush!

4.4.4 Migration of human populations

Migration is the movement of people from one place to another, usually over long distances, with the intention of settling in the new location. The map below summarises the early migration of humans from Africa into other continents.

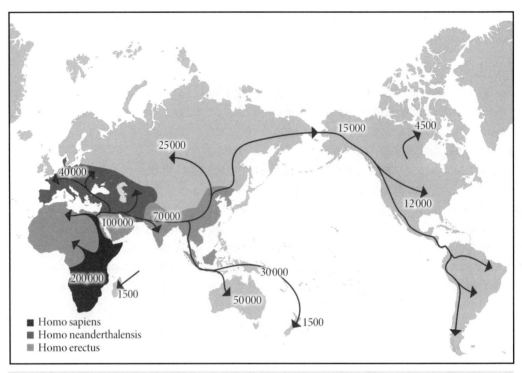

Map of early migration of humans

Mitochondrial DNA

Mitochondria have their own genome, consisting of approximately 16 500 base pairs. Mitochondrial DNA (mtDNA) is inherited only from the mother via the egg cell. All offspring have their mother's mtDNA. mtDNA contains 13 protein-coding genes and other genes that code for the production of tRNA and rRNA.

Studies of mtDNA are useful in evolution research for the following reasons.

- Each cell has many copies of mtDNA (unlike the single copy of nuclear DNA), making it more likely to survive for long periods (mtDNA samples from Neanderthals have been sequenced).

- There is no crossing over and recombination in mtDNA because it does not undergo meiosis, so there is no mixing of alleles from two parents.

- Mitochondrial DNA has a higher mutation rate than nuclear DNA, especially in a variable region of the genome known as the D-loop. This makes it useful in tracing divergence between populations over relatively short time periods.

- Variation in human mtDNA can indicate waves of migration and provide information on our early origins. Among modern human populations, various patterns of mutations in mtDNA can indicate where their ancestors came from, and how long ago they separated from other human populations. Mitochondrial DNA studies put the ancestor of modern humans at between 170 000 and 180 000 years ago in Africa.

Studies of mutations in mtDNA have been used to investigate the relationship between different populations of humans. **DNA sequencing** of mtDNA shows more variations in southern African populations than in populations from other parts of the world. More variation suggests a longer evolutionary history, supporting the hypothesis that the African population is the oldest human population. A study of genetic variation in a region of chromosome 12 also

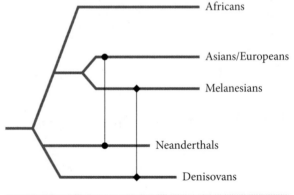

Interbreeding and gene exchange between modern humans

9780170479431

supports the 'out of Africa' hypothesis. The dating of fossils of modern humans also supports this hypothesis.

But things are never that simple. As modern humans spread throughout the world and encountered other hominin species, some interbreeding and gene exchange did occur. The following diagram shows where in our history this is thought to have occurred.

The migration of Aboriginal and Torres Strait Islander peoples

As seen in the previous map, modern humans reached Asia 70 000 years ago, before moving through South-east Asia and into Australia. *Homo erectus* has been in Asia for at least 1.5 million years and it is possible that *Homo sapiens* co-existed with *Homo erectus*; Indonesian *Homo erectus* findings suggest they survived there until as recently as 50 000 years ago. However, *Homo erectus* remains have never been found in Australia.

The Denisovans interbred with modern humans. Melanesians and Aboriginal Australians carry about 3–5% of Denisovan DNA. This can be explained through the interbreeding of eastern Eurasian Denisovans with the modern human ancestors of these populations as they migrated towards Australia and Papua New Guinea. Genomic evidence indicates that their ancestors arrived in the ancient landmass of Sahul (present-day New Guinea and Australia) approximately 55 000 years ago. This single founding event in Sahul was followed by a divergence of the Papuan and Aboriginal Australian ancestral populations and further genetic modification in the Aboriginal Australian population that could have occurred along with environmental change, such as desertification. Aboriginal Australians lived in a high level of isolation until only relatively recent times.

Sea levels rose about 8000 years ago, which cut off Australia from New Guinea and formed more than 250 islands in Torres Strait. There is archaeological evidence of dugong and turtle remains from approximately 7000 years ago, suggesting that humans have continually occupied the area.

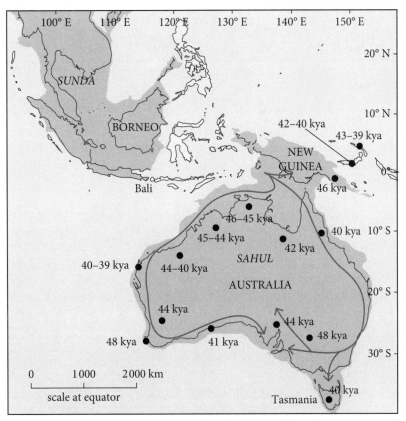

Migration routes taken by indigenous Australians. The dots indicate the archaeological sites and dates of the oldest found artefacts.

It is therefore understandable that Aboriginal and Torres Strait Islander peoples have a connection to Country and Place; a deep symbiotic relationship between the people and their lands and seas. Rather than owning the land, Aboriginal and Torres Strait Islander peoples belong to the land, and this relationship is continued by the environment, cultural knowledge and responsibility.

Aboriginal custom states that they are custodians of the land, and as such they take care of the land, which in turn provides for the people. Traditional methods for sustaining the land include backburning to prevent bushfires (and hence prevent food sources from being wiped out), predicting plant growth and encouraging animals to the land for hunting and propagation of plants.

These cultural practices are consistent with fossil evidence of extinct megafauna, such as *Genyornis newtoni*, a large flightless bird. The fossils show burn marks consistent with being cooked by humans.

Glossary

adaptive radiation When a single species diversifies relatively rapidly into many new species because of the availability of many different ecological niches

allele Different forms or variations of a gene

allele frequency The proportion of an allele at any particular gene locus in a population; ranges from 0 to 1

analogous structures Structures that have a similar function but different evolutionary origins

antigenic drift A random genetic mutation of a pathogen that causes minor changes in the antigen

antigenic shift A major change in the HA and/or NA proteins on a pathogen's membrane, which can result in new subtypes of influenza

binary fission The division of a cell into two without mitosis; the process by which a prokaryotic cell divides to form two cells

chromosomal mutations Changes to the structure of the chromosome as in block mutations, or changes to chromosome number as in polyploidy

common ancestor A species that represents a population of individuals from which other species of interest descended with adaptations

convergent evolution When unrelated species evolve similar adaptations in response to similar selection pressures

cross-protection When a host has immunity to one strain of virus that prevents infection by a closely related strain of that virus because their antigens are similar

deletion A mutation where a loss of genetic material has occurred

divergent evolution When members of a population develop adaptations to the different selection pressures over many successive generations

DNA sequencing The determination of the actual sequence of nucleotides in a DNA strand

duplication A mutation that occurs when one or more extra copies are made of a section of chromosome

embryonic development The growth and development of an embryo; also known as embryogenesis

equilibrium Balance between all acting influences

evolving Variation within a population that allows for some individuals to survive and reproduce better than others when there is a sudden change in conditions

extinction The permanent loss of a species; also applies to the loss of a population or higher taxon

(e.g. family)

fossil The preserved remains of an organism or direct evidence of its presence (e.g. footprints, casts or moulds)

fossil record The worldwide collection of fossils as they occur in the surface layers of Earth

founder effect Genetic drift that results from non-representative allele frequencies in a small founding population

frameshift A mutation where there has been a change in the DNA sequence that shifts the way a sequence is read

gene flow The transfer of alleles that results from emigration of individuals out of, and immigration of individuals into, populations

gene pool The range of genes and all their alleles present in a population

genetic bottleneck The loss of genetic variability when population size is severely reduced; results in the species being more vulnerable to selecting agents in the environment

genetic drift Changes in allele frequencies due to small population size and random causes

genome The total genetic material within a cell, an individual or a species

geographical barrier A canyon, river, cave, body of water or large expanse that physically separates one population from another

haemagglutinin (H) A glycoprotein found on the surface membranes of influenza viruses

half-life The time it takes for half of a sample of a radioactive isotope to decay to a more stable form

hominin Describes modern humans and extinct bipedal human-like hominids; includes the genera *Ardipithecus*, *Australopithecus* and *Homo*

hominoid A member of the superfamily *Hominoidea*; an ape, or tail-less primate

homologous structures Similar structures found across different species because they share a common ancestor; these structures may have different functions among the species

index fossil A fossil that is representative of a specific geological time

inheritance The passing on of genetic material from parents to offspring

inversion Rearrangement of chromosome segments, where the chromosome breaks, the broken section 'flips', and then reattaches

law of faunal succession Observations of the fossil record, where the sedimentary rock layers, or strata, contain fossils that supersede each other as you move up the layers

lysis The splitting of the membrane of a cell, which can be caused by chemical or physical means

macroevolution The evolution of species or high taxa (e.g. family)

mammal A warm-blooded vertebrate animal that has hair or fur, and the females secrete milk to nourish their offspring

microevolution The change in allele (and therefore phenotype) frequencies that occurs over time within a population and between populations of the same species

missense mutation A mutation that results in one amino acid being replaced by another amino acid in the encoded protein

multi-regional hypothesis The hypothesis that modern humans evolved from ancestral populations of other *Homo* species in many parts of the world and had recurrent gene flow

mutagen Any influence that increases the rate of mutation of DNA

mutation A change in DNA; either in a single nucleotide or in a large part of a chromosome

natural selection The process whereby individuals with certain heritable traits survive and reproduce more successfully than other individuals

neuraminidase (N) An enzyme found on the surface of influenza that allow the virus to be released from a host cell

nonsense mutation A mutation in which a codon for an amino acid is changed to one that codes for a stop codon, terminating translation

'out of Africa' hypothesis The hypothesis that modern humans originated only in Africa and migrated to all other parts of the world

peptidoglycan A substance that forms the cell walls in bacteria

phylogeny The evolutionary relationships that exist between species, often expressed as a tree-like diagram or represented by taxonomic classification

postzygotic A category of barriers in reproductive isolation that occurs after fertilisation has taken place, preventing survival of offspring

prezygotic A category of barriers in reproductive isolation that occurs before fertilisation can take place

primate A member of the order Primata; includes lemurs, lorises, tarsiers, monkeys, apes and modern humans

radiometric dating A method for determining the age of a rock or fossil based on the predictable rates of decay of naturally occurring radioactive isotopes present

random mating Mating where any egg has an equal chance of being fertilised by any sperm so that genetic variation is maintained

remains Parts that are left behind

reproductively isolated Where species cannot interbreed with related species due to a behavioural, physiological, genetic and geographical barrier

selecting agent *See* selection pressure

selection pressure A factor that acts to favour one phenotype over another; may be abiotic or biotic; also known as selecting agent

selective advantage Relative fitness of one phenotype over another

selective breeding Deliberate changing of the gene pool of a plant or animal species by controlling which population members breed

silent mutation A mutation in which the DNA codon for one amino acid becomes another DNA codon for the same amino acid; also referred to as a synonymous mutation

speciation The formation of two or more species from one ancestral population

species A group of similar organisms capable of breeding and exchanging genes with one another and whose offspring are capable of doing the same; also describes the lowest formal taxonomic rank and forms the second part of an organism's scientific name

transitional fossil A fossil that shows traits common to an ancestral group and a descendant group

translocation A mutation occurring when a section of one chromosome breaks off and reattaches to another chromosome

variation Any observed difference between two or more groups that can be caused by genetics or non-environmental factors

vestigial structures Structures that do not appear to have any function and may be left over from a past ancestor

9780170479431

Revision summary

In the table below, provide a brief definition of each of the key concepts. This will ensure that you have revised all the key knowledge in this Area of Study in preparation for the exam.

How are species related over time?	
Causes of changing allele frequencies in a population's gene pool	
Selection pressures as a source of new alleles	
Genetic drift, gene flow and mutations as a source of new alleles	
Biological consequences of changing allele frequencies in terms of genetic diversity	
Manipulation of gene pools through selective breeding programs	
Consequences of bacterial resistance and viral antigenic drift and shift in terms of ongoing challenges for treatment and vaccination	
Changes in species over geological time as evidenced from the fossil record	
Changes in species over geological time as evidenced from the fossil record: faunal (fossil) succession, index and transitional fossils, relative and absolute dating of fossils	
Evidence of speciation as a consequence of isolation and genetic divergence	
Allopatric speciation in Galapagos finches and sympatric speciation in *Howea* palms on Lord Howe Island	
Evidence of relatedness in species – structural morphology, homologous and vestigial structures, molecular homology (DNA and amino acids)	
The use and interpretation of phylogenetic trees	

››

››	How are species related over time?	
	Shared characteristics that define mammals, primates, hominoids and hominins	
	Evidence for major trends in hominin evolution from *Australopithecus* to *Homo*	
	The human fossil record as an example of a classification scheme that is open to differing interpretations that are contested, refined or replaced when challenged by new evidence, including evidence for interbreeding between *Homo sapiens* and *Homo neanderthalensis* and evidence of new putative *Homo* species	
	Ways of using fossil and DNA evidence (mtDNA and whole genomes) to explain the migration of modern human populations around the world, including the migration of Aboriginal and Torres Strait Islander populations and their connection to Country and Place	

Exam practice

Genetic changes in a population over time

Solutions start on page 241.

Multiple-choice questions

Question 1 ○●●●

A gene pool is

A the environment a species inhabits that includes a water source.

B the total sum of all of the alleles in a given population.

C all species that live in a certain area at the same time.

D all individuals of breeding age within a population.

Question 2 ○●●●

For the sequence of DNA below, what type of mutation has occurred?

Sequence: AGA CGG GAC TAT GAG TAG

Mutation: AGA CGG GAT CTA TGA GTA G

A Insertion

B Inversion

C Deletion

D Translocation

Question 3 ○○●●

Sickle cell anaemia is a blood disorder caused by a defective protein. The defect results from a change in a single amino acid. This change occurs because of a single base substitution in DNA. This is an example of a

A missense mutation.

B nonsense mutation.

C frameshift mutation.

D silent mutation.

Question 4 ○○●●

Which one of the following statements is NOT true of mutations?

A A mutation may affect future generations.

B A mutation may involve a change in gene expression.

C A mutation is usually of benefit to the organism that inherits it.

D A mutation can be brought about by environmental factors.

Question 5 ○○●●

An unusually high incidence of a particular genetic trait was found in a small, isolated population of bears. This is most likely due to

A genetic drift.

B divergent evolution.

C convergent evolution.

D speciation.

Question 6 ⬤⬤⬤

In an area of Indonesia, DDT was introduced in an attempt to control mosquito numbers. Widespread spraying of the insects took place over a number of years. During this time, samples of the mosquito population were collected and tested for their susceptibility to DDT. The results are shown in the table below.

Death rate of mosquitos sprayed with DDT	
Year	% killed
1	100
2	100
3	93
4	86
5	50
6	32

On the basis of the results obtained, it is reasonable to conclude that

A the introduction of DDT caused a mutation that made the mosquitos resistant to DDT.

B over a period of time, individual mosquitos developed a tolerance to DDT.

C DDT provided a selection pressure on the mosquito population, leading to a change in the frequency of alleles for DDT resistance.

D the mosquito population at the end of the sixth year is likely to be resistant to a variety of insecticides.

Question 7 ©VCAA VCAA 2011 (2) SA Q20 (adapted) ⬤⬤⬤

The soapberry bug (*Jadera haematoloma*) uses its long beak to penetrate the fleshy fruit of plants to feed on the seeds at the centre. The bug feeds on the native soapberry tree. The bug also feeds on the fruit of the introduced golden rain tree. Investigators measured the average beak length of the soapberry bug over 80 years. The results are shown below.

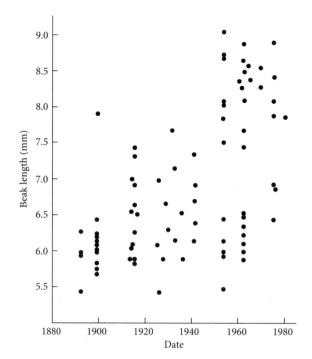

In neighbouring regions, the fruits of other introduced plants have also been used by the soapberry bug. Male and female soapberry bugs from different regions can interbreed. Evidence indicates that genetic isolation of some of these populations is gradually occurring.

The situation that would lead to an increase in genetic isolation would be if

A different types of host plants have fruiting seasons that do not overlap.

B pheromones of female soapberry bugs attract soapberry males from neighbouring populations.

C the soapberry tree is common throughout the distribution and each tree produces large amounts of fruit.

D male soapberry bugs new to a region are reproductively active, whereas female bugs need to feed before becoming reproductively active.

Question 8 ©VCAA VCAA 2019 SA Q26 ●●●

Consider the diagram at the right showing the gene pool of a population over 20 generations.

It would be correct to conclude that, over the 20 generations

A genetic diversity is increasing in this population.

B individuals with the genotype RR had a selective advantage in this population.

C the frequency of each allele is equal in Generation 1 but not in other generations.

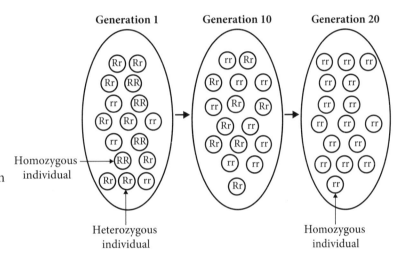

D new advantageous alleles for this gene were introduced as individuals joined this population.

Question 9 ●●

Cheetahs have long been desired by hunters for their pelts. As a result, the population of cheetahs has been significantly reduced to the point that they are now listed as vulnerable, with some scientists declaring they should be categorised as endangered. With this significant decline in wild cheetah numbers, what might be the consequence to the species?

A The species is evolving to become faster and more agile.

B Cheetahs have less space to roam and find potential mates.

C Hunters are finding it difficult to locate them as they are in such low numbers.

D They are vulnerable to disease due to a decline in genetic variation.

Question 10 ©VCAA VCAA 2016 SA Q37 ●●

Tiburon is an isolated island off the coast of Mexico. Desert bighorn sheep became extinct on this island hundreds of years ago. In 1975, 20 desert bighorn sheep were taken from a population in the American state of Arizona (shown on the map below) and were reintroduced to Tiburon Island. By 1999, the population of desert bighorn sheep on Tiburon Island had risen to 650.

Which one of the following statements about this 1999 population of desert bighorn sheep on Tiburon Island is correct?

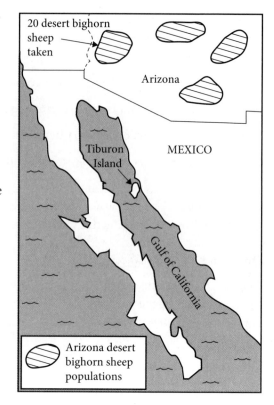

A The gene pool of this population will be identical to the gene pool of the Arizona populations.

B This population has less genetic variation than the Arizona populations and is an example of the founder effect.

C This population will have become a new species because the mutation rate on Tiburon Island will be much higher than in Arizona.

D Having been through a population bottleneck, the current population will now show increased genetic variation compared to the Arizona populations.

Question 11 ◖◗◗

Australia's dairy industry generates $13 billion annually. Farmers aim to improve their herds by breeding cows that produce more milk with a higher protein and fat content, while consuming fewer calories (grass and grain). They do this by choosing bulls with desirable characteristics to breed with their most productive cows. Over many generations, a herd increases its production of milk high in protein and fat. This is an example of

A natural selection.

B artificial selection.

C evolution.

D the bottleneck effect.

Question 12 ◖◗◗◗

When a doctor prescribes medications for a bacterial infection, they explicitly ask that the patient takes the whole course of medication even if they are feeling better. This is because the emergence of antibiotic-resistant diseases in humans is increasing. This means that

A the most sensitive bacteria are killed first, followed by the more resistant bacteria.

B viruses are becoming resistant to antibiotics.

C antibiotics are causing mutations in bacteria resulting in resistance.

D humans cells are becoming sensitive to the antibiotic.

Question 13 ◖◗◗◗

Cleanliness is imperative in hospitals to prevent the spread of disease to susceptible people. Patients may be admitted to hospital if general antibiotics are not effective and intravenous antibiotics are required. An implication of treating an antibiotic-resistant strain of bacteria in a hospital setting is that

A the length of time a patient stays in hospital for such an infection would be shorter.

B people with antibiotic-resistant bacterial infections will need to be treated with antiviral drugs instead.

C the resistant bacteria can become life threatening to people already ill in hospital if the bacteria are not contained.

D the patient develops a tolerance to the antibiotic and it therefore no longer treats the infection.

Short-answer questions

Question 14 (5 marks) ☐☐☐

An entomologist trekked into the Australian rainforest to observe the rhinoceros beetle. She measured the size and mating rituals, as well as their diet. Twenty years later she trekked to the same geographical location and found that all the beetles had slightly increased in size compared to the population of beetles she had earlier encountered.

a What term describes this minor change in the population of the beetles?   1 mark

b How might this change have occurred? Give an example.   2 marks

c What is the difference between gene flow and genetic drift? Describe this using
the example above.   2 marks

Question 15 (7 marks) ☐☐

The following table shows a strand of template DNA in its original form, and the same strand after it has undergone various mutations.

Original DNA	AAT CGG TAG
Mutation 1	AAC CGG TAG
Mutation 2	AAT TCG GTA G
Mutation 3	AAT CGG TTG

a What would be the base sequence in a complementary mRNA strand formed from
the original DNA?   1 mark

b **i** Describe the mutation that has occurred in Mutation 1.   1 mark

   ii What would be the base sequence in a complementary mRNA strand formed from
the DNA shown in Mutation 1?   1 mark

   iii Use the codon table below to determine what effect Mutation 1 will have on the
peptide chain coded for by this DNA segment.   2 marks

Second base

	U	C	A	G	
U	UUU ⎤ Phe UUC ⎦ UUA ⎤ Leu UUG ⎦	UCU ⎤ UCC ⎥ Ser UCA ⎥ UCG ⎦	UAU ⎤ Tyr UAC ⎦ UAA Stop UAG Stop	UGU ⎤ Cys UGC ⎦ UGA Stop UGG Trp	U C A G
C	CUU ⎤ CUC ⎥ Leu CUA ⎥ CUG ⎦	CCU ⎤ CCC ⎥ Pro CCA ⎥ CCG ⎦	CAU ⎤ His CAC ⎦ CAA ⎤ Gln CAG ⎦	CGU ⎤ CGC ⎥ Arg CGA ⎥ CGG ⎦	U C A G
A	AUU ⎤ AUC ⎥ Ile AUA ⎦ AUG Met/ Start	ACU ⎤ ACC ⎥ Thr ACA ⎥ ACG ⎦	AAU ⎤ Asn AAC ⎦ AAA ⎤ Lys AAG ⎦	AGU ⎤ Ser AGC ⎦ AGA ⎤ Arg AGG ⎦	U C A G
G	GUU ⎤ GUC ⎥ Val GUA ⎥ GUG ⎦	GCU ⎤ GCC ⎥ Ala GCA ⎥ GCG ⎦	GAU ⎤ Asp GAC ⎦ GAA ⎤ Glu GAG ⎦	GGU ⎤ GGC ⎥ Gly GGA ⎥ GGG ⎦	U C A G

First base (left) / Third base (right)

c Which of the mutations (1–3) will result in a frameshift in the genetic code? What will be the
consequence of this?   2 marks

Question 16 (6 marks) ⬤⬤⬜

Ellis-van Creveld syndrome is a rare genetic condition first discovered in 1940. Sufferers are of equal sex distribution and of normal intelligence. All, however, are moderately to severely dwarfed and have malformations of the hands, including polydactyly (extra fingers). There are also characteristic facial malformations, including a short upper lip and dental problems. In 50% of cases there are also cardiovascular problems. It was these cardiovascular problems that led to a high mortality rates during childhood. In 1954, researchers established that this condition was autosomal and recessive.

In the 1960s, more cases of Ellis-van Creveld syndrome were found in Lancaster County in the USA than had been reported in the world to that date. All these cases were found among the Amish community. The ancestors of today's Amish fled religious persecution in Switzerland between 1720 and 1770. In all, some 200 people settled in Lancaster County and remained reproductively isolated from the rest of the American population. There are more than 7000 Amish today in this area.

Of the initial settlers, either one or two individuals were carriers of the allele for Ellis-van Creveld syndrome. Today, 7% of the Amish community carry at least one copy of this allele, compared with 0.1% in the general population. This is a clear example of the founder effect in a human population.

a i What is meant by the term 'founder effect'? 1 mark

   ii Explain how the founder effect has influenced gene frequency in the Amish population. 2 marks

b Remaining reproductively isolated has resulted in a lot of inbreeding in the Amish population of Lancaster County. How has inbreeding influenced the prevalence of this disease? 1 mark

c As stated above, there were one or two carriers in the initial population of 200. This means the original gene frequency would have been 0.5 or 1%. How do you account for the fact that it has risen to 7%, despite the obvious negative effects of this condition? 1 mark

d Until recently, Ellis-van Creveld syndrome was often fatal in childhood due to the number of cases in which there were cardiovascular problems. This condition is now treatable, and many more affected children are surviving well into adulthood. What effect is this likely to have on the future population of the Lancaster County Amish? 1 mark

Question 17 (7 marks) ©VCAA VCAA 2018 SB Q7 ⬤⬤⬤

Populations of the lizard species *Anolis sagrei* are found on the many islands of the Bahamas. There is natural variation between the phenotypes of individuals within each population.

a Explain how natural variation can exist between individuals within a lizard population. 3 marks

In 2004 a hurricane killed all populations of *A. sagrei* lizards on seven of the smaller islands. Scientists randomly chose seven males and seven females from a remaining population on a large island. They introduced one male and one female to each of the seven smaller islands. Over the next three years, the scientists noted that the size of the populations increased on each of the seven smaller islands. The scientists measured the genetic diversity within each of the populations and found there was lower genetic diversity in each new population compared with the population on the large island.

b Explain the reasons for the lower genetic diversity of the new populations on the smaller islands compared with the population on the large island. 2 marks

c The scientists noted that, after three years, there was a significant decrease in the average length of the hind legs of the lizards living on the smaller islands compared with those on the large island. Explain what may have happened on the smaller islands to produce this decrease in the average length of hind legs. 2 marks

Question 18 (6 marks)

Rabbits were introduced into Australia from Europe in the 1800s and quickly spread across the continent. They are highly adaptable to new environments because all they need is soil for burrowing and short grasses for grazing. Rabbits have even spread into desert landscapes. Their spread across the continent has been aided by their ability to have multiple offspring, multiple times a year.

a What is the only way new alleles can be formed?   1 mark

b Using the information above, how might allele frequency be able to change quite rapidly in rabbit populations?   1 mark

c Explain the process of how rabbits became adapted to the hotter climates of the desert with regards to allele frequencies.   4 marks

Question 19 (12 marks)

a The characteristics of a species can be changed in three main ways: natural selection, selective breeding and genetic engineering. Explain the process and provide an example of

   i natural selection.   2 marks

   ii selective breeding.   2 marks

   iii genetic engineering.   2 marks

b Genetic engineering can produce *genetically modified organisms* and/or *transgenic organisms*. Distinguish between these terms.   2 marks

c List **two** advantages and two disadvantages of selective breeding.   4 marks

Question 20 (11 marks)

African sleeping sickness, or trypanosomiasis, is a disease caused by a protozoan microorganism of the genus *Trypanosoma*. The disease is transmitted by the tsetse fly and hence occurs only in tropical regions where these flies breed. *Trypanosoma* also infects domestic cattle.

The disease passes through three phases in infected people. Initially there is localised pain, inflammation and itching at the site of the infection. This is followed by systemic infection causing fever, chills, fluid build-up and enlarged lymph glands. Finally, the organism invades the brain and spinal cord, causing sleepiness, coma and, if untreated, death. It may take several years for the disease to reach this final phase.

a What process is responsible for the initial localised reaction to infection by *Trypanosoma*?   1 mark

b *Trypanosoma* avoids complete destruction by the host's immune system by antigenic variation. At present there are no vaccines to protect against infection by *Trypanosoma*.

   i How does antigenic variation enable the parasite to survive?   1 mark

   ii Suggest why attempts to develop a vaccine have been unsuccessful.   3 marks

   iii Suggest another possible means to control the spread of trypanosomiasis.   1 mark

Like trypanosomiasis, many other diseases transmitted by insects are mild in their initial effects, often taking years to reach a fatal phase of the disease. Where the transmission of a disease is more direct, the progress is usually more rapid (see table below).

Mode of transmission	Example	Time taken for the disease to reach its most serious stage
Airborne	Cold, flu (viral)	Several days to weeks
Direct transmission	Gonorrhea (bacterial)	Several weeks to months
Insect transmission	Trypanosomiasis (protozoan)	Months to years

c What advantage is there for pathogens carried by insects to take several months to years to produce life-threatening symptoms in the hosts? 1 mark

d For which condition or conditions shown in the table above would antibiotics be an effective treatment? Explain your choice(s). 2 marks

e Doctors recommend that patients take all of the antibiotics prescribed even if symptoms have subsided. Why is this important? 2 marks

Question 21 (9 marks) ⬤⬤⬤

In 2009, swine flu was first detected in Mexico. The virus was identified as an influenza virus of type H1N1. Like the global flu pandemic of 1918–19, this H1N1 flu virus appeared to be particularly harmful to otherwise healthy young adults and children. Despite attempts by the Mexican authorities to contain the outbreak, it soon grew to epidemic proportions and was later declared a pandemic.

a What does the type name H1N1 refer to? 2 marks

b What is the difference between antigenic drift and antigenic shift? 2 marks

c Why would swine flu have been categorised as a pandemic? 1 mark

d At the time of the Australian outbreak, many people were warning against eating pork and pork products. Explain whether you believe this was a sensible precaution to take to contain the spread of swine flu. 1 mark

Relenza is a commercially synthesised drug, taken by inhalation, designed to combat influenza by disrupting the action of a viral enzyme. It competitively inhibits the neuraminidase active site. This blocks the virions from being released from an infected cell.

e Outline **two** reasons why inhalation could be a more effective means of administering the drug than ingestion. 2 marks

f Why is it important to vaccinate against influenza viruses annually? 1 mark

Changes in species over time

Solutions start on page 245.

Multiple-choice questions

Question 1 ●○○

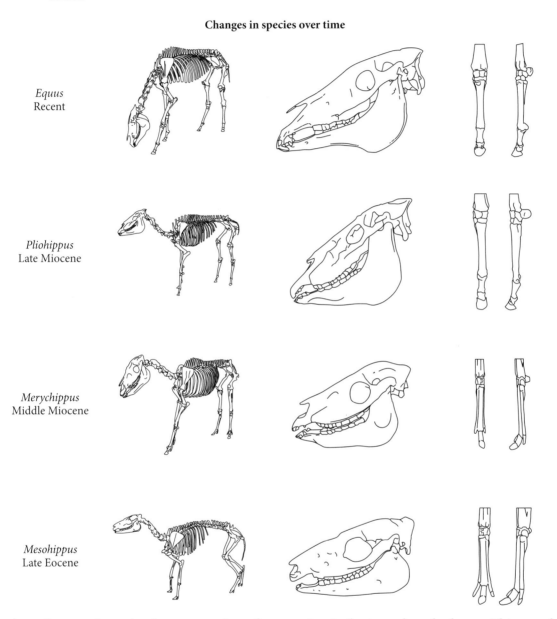

Changes in species over time

The above diagram shows the changes over time of one species similar to modern-day horses. This record shows

A index fossils.

B trace fossils.

C transitional fossils.

D incomplete fossils.

Question 2 ©VCAA VCAA 2016 SA Q38 ●●●

In India, a group of scientists was studying fossils from a coal deposit formed during the Permian period (290–245 million years ago). They found three fossil species from the same genus in different levels (strata) of the coal. When radiocarbon dating on these fossils was performed, it showed exactly the same levels of carbon-14 in all three fossil species. The data is summarised in the table below.

Fossil species	Depth at which fossil was found in the coal deposit (m)	Proportion of carbon-14 (%)
Gangamopteris major	6.2	0.0001
Gangamopteris obliqua	8.1	0.0001
Gangamopteris clarkeana	4.7	0.0001

Which one of the following is the correct conclusion to draw from these findings?

A There is no evolutionary relationship between these three fossil species.

B *G. clarkeana* is the common evolutionary ancestor of *G. major* and *G. obliqua*.

C As carbon dating is a more reliable dating technique than analysis of strata in coal deposits, the fossils of *G. major*, *G. obliqua* and *G. clarkeana* are all of the same age.

D An analysis of strata in coal deposits is a more reliable dating technique than carbon dating for Permian fossils; the fossil of *G. major* is younger than the fossil of *G. obliqua*.

Question 3 ●●●

The half-life of carbon-14 is 5730 years. The carbon-14 become nitrogen-14 in the decaying process. A fossil was recently uncovered that had a carbon:nitrogen ratio of 1:3. The scientists involved estimated the age of the fossil to be around 17 200 years old. Are the scientists correct?

A Yes, because the scientists found the fossil in strata with index fossils.

B Yes, because the fossil has undergone 3 half-lives, resulting in the ratio 1:3.

C No, because the fossil is too old to be dated with carbon dating. Potassium argon dating would be a better method.

D No, because the fossil has undergone 3 half-lives, resulting in the ratio 3:1.

Question 4 ●○○

Speciation is

A when two individuals from a population can produce viable offspring.

B when a species exists in isolation from all other species and can flourish in the local environment.

C when individuals from different species can produce offspring but they are infertile.

D the evolutionary process where populations evolve to become distinct species.

Question 5 ●●○

Which of the following examples would lead to allopatric speciation?

A Some members of a fruit fly population being swept away in a storm onto another island, from which they could not return due to the distance travelled

B All members of a mountain goat population being isolated on an outcrop due to an earthquake that has redirected a river

C Some members of the hawthorn fly population choosing to lay their eggs in green apples and not red apples

D The damming of a river by beavers so fish could not swim back up to their breeding grounds

Question 6

Which of the following is an example of a prezygotic reproductive barrier?

A Two species of foxes produce infertile offspring.

B Two related plants in the same area have different flowering seasons.

C A hybrid embryo of two species fails to develop.

D A hybrid offspring of two related bird species dies soon after birth.

Short-answer questions

Question 7 (8 marks)

The picture below shows soil investigations at three different sites within a 100 km radius of each other.

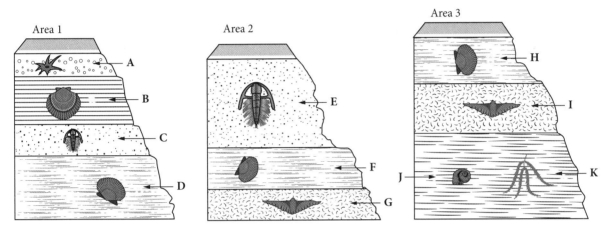

a The diagrams indicate different soil layers. What are these called and how are they formed?   2 marks

b Would Organism A and Organism E have been alive at the same time? Use evidence from the diagrams in your answer.   1 mark

c Label the layers on the three diagrams 1–6, with 1 being the oldest layers with the oldest fossils and 6 being the youngest layers with the youngest fossils.   2 marks

d Organism D has been identified as an index fossil. What does that mean?   1 mark

e Would the fossils found in these layers be a complete set of all of the creatures living in this area over a given time frame? Why/why not?   2 marks

Question 8 (12 marks)

Two species of Eucalypts are *Eucalyptus viminalis* and *Eucalyptus miniata*.

E. viminalis is commonly known as the manna gum, white gum or ribbon gum, and is endemic to south-eastern Australia. It has small cream-coloured flowers in clusters of three that are pollinated by insects during the summer months.

E. miniata is commonly known as the Darwin woollybutt or woolewoorrng, and is a species of medium-sized to tall tree that is endemic to northern Australia. It has rough, fibrous, brownish bark on the trunk, and smooth greyish bark above. It produces bright orange or scarlet flowers that are pollinated by birds in the winter months.

a Describe how *E. miniata* and *E. viminalis* became different species.   4 marks

b Sometimes different species can produce hybrid offspring. In the natural world, can *E. miniata* and *E. viminalis* cross pollinate? Give **two** pieces of evidence to support your answer.   2 marks

c A Eucalypt forest of 2000 acres was cleared to make way for farming land, leaving only 200 acres of natural forest remaining. How might this be an example of the bottleneck effect?   3 marks

The oldest known examples of eucalypt fossils are 52-million-year-old flowers, fruits and leaves found in South America.

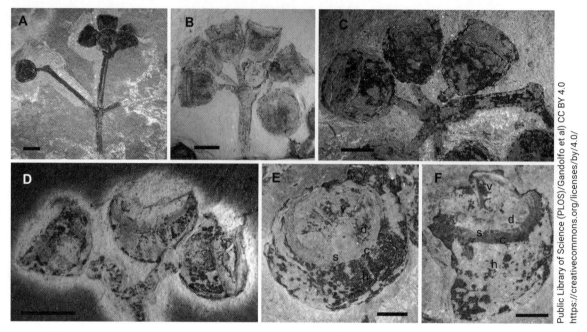

d Would these fossils have been dated using radiometric dating such as carbon dating? Why/why not?

1 mark

e Describe how these fruits could have become fossils.

2 marks

Question 9 (6 marks)

In the 1800s, Charles Darwin formed a theory with regards to 13 breeds of finches on the different islands of the Galapagos.

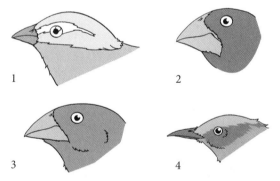

1 *Geospiza parvula*   2 *G. magnirostrus*   3 *G. fortis*   4 *Certhidea olivacea*

a What did Darwin hypothesise regarding the different breeds of finches?

1 mark

b Darwin looked at **three** traits within the finches. What were they?

1 mark

c What did Darwin believe influenced these traits on each of the different islands?

1 mark

d How did this result in finches that look remarkably different to the finches on mainland South America?

2 marks

e What is the name of the theory that resulted from this discovery?

1 mark

U4 – AOS2 – EXAM PRACTICE

Question 10 (7 marks) ©VCAA VCAA 2016 SB Q9 ●●●

Galapagos tortoises (*Chelonoidis spp.*) can be found on many of the islands that make up the Galapagos Islands. Originally, 14 different species were identified based on the islands on which they lived and on their morphology.

Santa Cruz, the second largest of the Galapagos Islands, has two isolated tortoise populations. Population A contains more than 2000 individuals covering an area of 156 square kilometres. Population B is a small population of 250 individuals covering an area of 40 square kilometres.

The position of the two populations on the island of Santa Cruz is shown at the right. The two populations are separated by a distance of 20 kilometres.

In 2015, scientists investigated whether the individuals of the two populations belong to the same species or whether they are two different species. Average measurements of skull size were calculated for tortoises belonging to both populations A and B. The skulls were measured in six different places. The six measurements were also compared to average measurements taken from skulls of other Galapagos tortoise species. The results are shown in the table below. Comparisons have been made with three other Galapagos tortoise species.

Measurement position	Average skull measurement				
	Population B	Population A	*Chelonoidis vicina*	*Chelonoidis chathamensis*	*Chelonoidis ephippium*
1	118	98	86	80	74
2	40	37	28	27	25
3	21	18	16	14	12
4	26	23	21	18	17
5	10	9	8	7	6
6	19	17	16	14	13

a Consider the data given. Does the data support the hypothesis that individuals in Population A belong to a different species from individuals in Population B? Explain your answer. 2 marks

b Scientists have carried out genetic studies on the two populations. Give an example of genetic evidence that may be produced by scientists to support the hypothesis that individuals of the two populations belong to different species. Explain your answer. 2 marks

c Some scientists thought that allopatric speciation may have occurred on the island of Santa Cruz.

i Name a feature that scientists would look for in the island environment to support the occurrence of allopatric speciation. 1 mark

ii Explain how the feature named in part **c i** could contribute to allopatric speciation. 2 marks

Question 11 (3 marks)

Lord Howe Island off the coast of Australia is characterised by subtropical forests. These forests contain a variety of palm species. Some palms survive better in the volcanic acidic soil, such as the *Howea belmoreana*; others thrive in the calcareous alkaline soil, such as *Howea forsteriana*. Growing in calcarenite soil is stressful for plants because such soil is poor in nutrients. This physiological stress has caused a shift in the way in which the flowers of *H. forsteriana* mature. Researchers have noted that, when found growing on richer volcanic soils, the flowers mature in a way that is synchronous, where most of the population of plants flower/produce and receive pollen at the same time, not unlike the flowers of *H. belmoreana*. It has been identified that many of these palm species evolved from a common ancestor.

a Describe the process of evolution involving the different palm species on the island. 2 marks

b What is the name given to this process? 1 mark

Determining the relatedness of species

Solutions start on page 248.

Multiple-choice questions

Question 1

Comparison of the arm of a human and the flipper of a seal shows similar underlying bone structure. This is an example of

A homology.

C convergence.

B analogy.

D symmetry.

Question 2

The diagram shows the early embryonic development of a chicken and a rabbit.

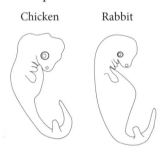

Chicken   Rabbit

What is the most reasonable explanation for the presence of gill slits and tails in these embryos?

A Chickens and rabbits share a very recent common ancestor.

B Ancestral vertebrates had the genes to produce gill slits and tails, and these remain active in embryonic development in both these species.

C It is an example of convergent evolution caused by similar embryonic environments.

D These species have evolved from modern fish and so possess fish-like structures.

Question 3

Old World and New World vultures both have strong, hooked beaks and naked heads adapted to probing animal carcasses for food. However, evidence from biochemical analysis shows that New World vultures are more closely related to storks than to Old World vultures. Structural similarities between New World and Old World vultures are due to

A divergent evolution.

C genetic drift.

B convergent evolution.

D migration.

Question 4 ○●●

The diagram represents the relationship between several species.

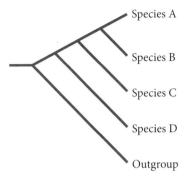

Which of the following is true of the relationships between these species?

A The DNA base sequences in species A and D would be most similar.

B The two species A and B share the most recent common ancestor.

C Species C is an ancestor of both species A and B.

D All four species, A to D, share the same mitochondrial DNA.

Use the following information to answer Questions 5 and 6.

Consider the following phylogenetic tree, which summarises the evolutionary relationships between certain fish species.

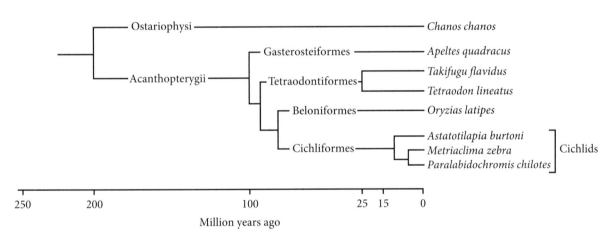

Question 5 ©VCAA VCAA 2019 SA Q24 ○●●

O. latipes is most closely related to

A *A. quadracus.*

B *T. lineatus.*

C *P. chilotes.*

D *T. flavidus.*

Question 6 ©VCAA VCAA 2019 SA Q25 ○●●

Which one of the following statements is correct?

A Cichlids diverged to form three distinct species 100 million years ago.

B *C. chanos* was the last species to diverge from the most distant common ancestor.

C Gasterosteiformes, Beloniformes and Cichliformes do not share a common ancestor.

D *T. flavidus* and *T. lineatus* diverged to form two distinct species 25 million years ago.

Short-answer questions

Question 7 (9 marks)

Below is a phylogenetic tree of some common Australian animals.

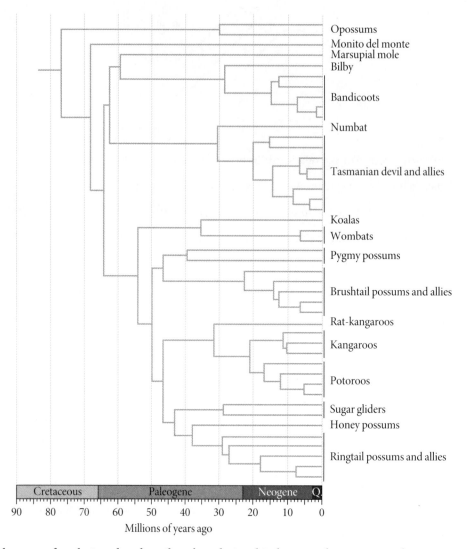

a Name the type of evolution that describes the relationship between kangaroos and potoroos.   1 mark

b How long ago did koalas and wombats diverge from a common ancestor?   1 mark

c Which species diverged from a common ancestor around 60 million years ago?   1 mark

The diagram below shows the hindfeet and claws of some Australian native species.

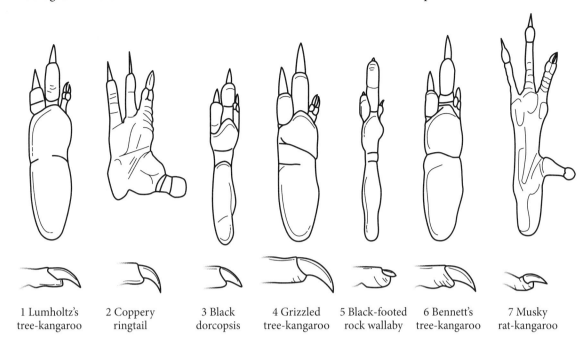

| 1 Lumholtz's tree-kangaroo | 2 Coppery ringtail | 3 Black dorcopsis | 4 Grizzled tree-kangaroo | 5 Black-footed rock wallaby | 6 Bennett's tree-kangaroo | 7 Musky rat-kangaroo |

d **i** Using just the hind foot and claw images, create a phylogenetic tree to show the evolutionary relationships between the animals.   3 marks

   ii Explain the reasoning behind your response to part **d i**.   3 marks

Question 8 (6 marks) ●●●

A fossil was recently found that showed physical features similar to a modern-day ostrich. However, DNA hybridisation techniques indicate that the fossil is more closely related to a crane than to an ostrich.

a Outline the steps involved in the DNA hybridisation technique following the extraction and cutting of DNA.   3 marks

b What would the results obtained look like in this scenario?   3 marks

Question 9 (12 marks) ©VCAA VCAA 2018 SA Q9 ●●

Cetaceans (whales, porpoises and dolphins) are marine mammals belonging to the order *Artiodactyla* (even-toed hoofed mammals). The closest living relative of cetaceans is the hippopotamus.

Phylogenetic tree A summarises the evolutionary relationships of four present-day cetacean species and the hippopotamus.

Phylogenetic tree A

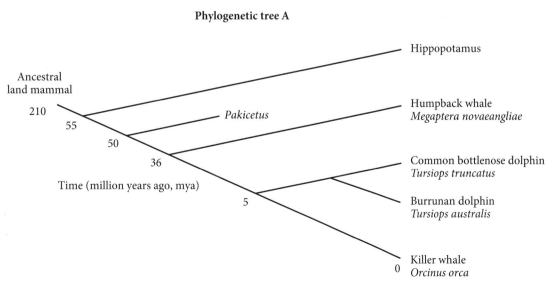

a   What does the length of the line that represents the evolution of *Pakicetus* suggest?   1 mark

b   A fossil named *Ambulocetus* was found in 1992 and dated at 49 million years old. Some palaeontologists believe that it is a transitional fossil between the ancestral land mammal shown in Phylogenetic tree A and present-day cetaceans.

Predict **two** structural features of the *Ambulocetus* fossil that would provide evidence to support the hypothesis that it is a transitional fossil and suggest a survival advantage of each feature.   3 marks

Sharks are marine fish of the order *Chondrichthyes* (cartilaginous fish). Phylogenetic tree B summarises the evolutionary relationships of three present-day shark species and main fish classes.

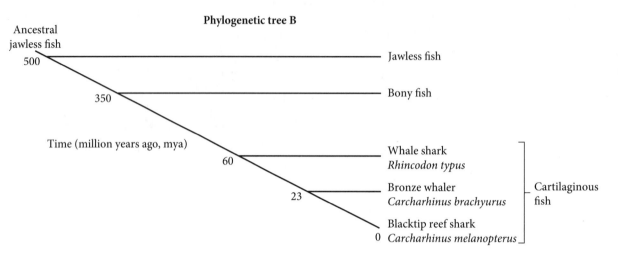

Phylogenetic tree B

The table provides information about the present-day species shown in phylogenetic trees A and B.

Marine animal		Average length	Diet
	Humpback whale *Megaptera novaeangliae*	16	Filters plankton and krill by sucking water into mouth through baleen plates
	Common bottlenose dolphin *Tursiops truncatus*	4	Hunts fish, squid and crustaceans
	Burranan dolphin *Tursiops australis*	3	Hunts fish and squid
	Killer whale *Orcinus orca*	10	Hunts sea birds, squid, seals, baleen whales, dolphins, fish, sharks and sea turtles depending on location
	Whale shark *Rhincodon typus*	12	Filters plankton, small squid and fish through filter pads and 300 rows of teeth
	Bronze whaler *Carcharhinus brachyurus*	3	Hunts squid, bony fish and other cartilaginous fish
	Blacktip reef shark *Carcharhinus melanopterus*	2	Hunts small bony fish, squid and shrimp

c **i** Give a specific example of divergent evolution using two animals from the
information provided, and justify your response.   2 marks

ii Give a specific example of convergent evolution using two animals from the
information provided, and justify your response.   2 marks

In 2011, Australian scientists identified the burrunan dolphin as a separate species from the common
bottlenose dolphin. They gathered evidence from living dolphin populations as well as museum
specimens.

d Briefly describe **two** types of evidence that the scientists would have used to establish
whether the burrunan dolphin is a separate species from the common bottlenose dolphin.   2 marks

e Burrunan dolphins are found only in Port Phillip Bay and the Gippsland Lakes in Victoria. There are
only 150 burrunan dolphins alive today. The Port Phillip Bay population is very isolated and rarely
mixes with dolphins outside the bay. Port Phillip Bay is impacted by the human population and
industry of Melbourne and surrounding towns, and is used heavily for recreation, fishing and
shipping. Suggest **two** possible future outcomes for the Port Phillip Bay population of burrunan
dolphins. Justify each outcome.   2 marks

Human change over time

Solutions start on page 250.

Multiple-choice questions

Question 1

An outgroup is a species or group of species that is related to, but does not belong to, the group shown in
the diagram.

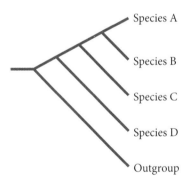

If species A to D in this diagram represent hominoids, then the outgroup could be

A tree shrews.

B gorillas.

C birds.

D baboons.

Question 2

Which of the following features distinguishes primates from other mammals?

A Tail and forward-facing eyes

B Flat nails and dextrous hands

C Arms longer than legs and even-sized teeth

D Fur and bipedal stance

Question 3 ●●●

The earliest hominins differed from apes that existed at the same time in that they

A had significantly larger brains.

B had less body hair.

C were bipedal.

D lived in family groups.

Question 4 ●○○

Members of the species *Homo neanderthalensis*

A had smaller brains than modern humans.

B inhabited wide areas of the African continent.

C are ancestors of modern humans.

D are more closely related to modern humans than to modern apes.

Question 5 ©VCAA VCAA 2018 SA Q38 ●●○

Consider the evolution of hominins. Which one of the following statements about hominin evolution is correct?

A *Homo sapiens* and *Homo neanderthalensis* are the only present-day hominin species.

B Members of the *Australopithecus* genus are not classified as hominins.

C *Homo erectus* was a bipedal primate.

D All hominoids are also hominins.

Question 6 ©VCAA VCAA 2015 SA Q37 ●●○

Below are three images of fossil hominin skulls. Which sequence best shows the order from the most ancient fossil skull to the most modern fossil skull?

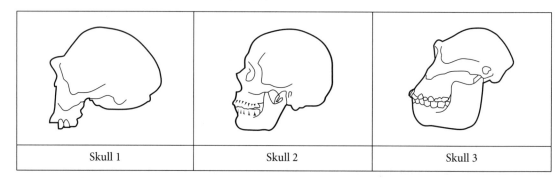

| Skull 1 | Skull 2 | Skull 3 |

A Skull 3, Skull 1, Skull 2

B Skull 1, Skull 2, Skull 3

C Skull 3, Skull 2, Skull 1

D Skull 1, Skull 3, Skull 2

Question 7 ©VCAA VCAA 2019 SA Q33 ●●○

Which group of characteristics best reflects the trends in hominin evolution from the *Australopithecus* species to the *Homo* species?

A Decreasing tooth size, increasing size of brow ridges, increasingly bowl-shaped pelvis, increasing size of zygomatic arch

B Decreasing tooth size, decreasing size of brow ridges, decreasing arch of feet, more-opposable big toe

C Increasing jaw size, decreasing size of zygomatic arch, increasing arch of feet, decreasing tooth size

D Decreasing size of canines, decreasing size of zygomatic arch, increasingly bowl-shaped pelvis, increasing arch of feet

Question 8 ○●●

People from parts of Melanesia have around 6% Denisovan DNA in their genomes. A possible explanation for this is that

A DNA samples from the fossils found are so old that the sequencing was incomplete.

B Denisovans migrated from Africa earlier than once thought.

C *Homo sapiens* are descended from Denisovans.

D Homo *sapiens* interbred with Denisovans at some point during the evolution of modern humans.

Question 9 ©VCAA VCAA 2017 SA Q35 ●●●

Modern African *Homo sapiens* do not contain Neanderthal DNA. Modern non-African *H. sapiens* contain a small percentage of Neanderthal DNA because of interbreeding between Neanderthals and *H. sapiens*. This interbreeding is thought to have occurred within the time period 65 000 to 47 000 years ago. A recent study has found *H. sapiens* DNA in the genomes of 100 000-year-old Neanderthal remains. From this new discovery, it would be reasonable to conclude that

A modern Africans are the descendants of Neanderthals.

B there was an early migration of *H. sapiens* out of Africa before 100 000 years ago.

C the ancestors of modern Africans migrated from Europe to Africa between 65 000 and 47 000 years ago.

D approximately 100 000 years ago, Neanderthals bred with *H. sapiens* in Africa before the Neanderthals spread to the rest of the world.

Question 10 ○●●

The oldest known fossil attributed to our genus, *Homo*, dates to about 2.8 million years ago and was reported in 2015. The ability to make stone tools was once thought to be the hallmark of the *Homo* species, but the oldest stone tools are now thought to be 3.3 million years old. What could be the explanation for this?

A Scientists have not yet found the early form of *Homo* that made the stone tools.

B The dating method used was not specific enough.

C Stone tools were only made by *Homo* species.

D Australopithecines were unable to make them because they did not have prehensile hands to grasp them.

Question 11 ○●●

Differences in mitochondrial DNA can be used to determine species relatedness. This technique would be most useful in determining the relatedness of

A mosses and grasses.

B cyanobacteria and other bacteria.

C fungi and animals.

D chimpanzees and humans.

Question 12 ⬤⬤

The following table shows the percentage difference in DNA sequences between humans and three other primate species.

Groups compared	% difference in DNA sequences
Human/chimpanzee	1.6
Human/gibbon	3.5
Human/rhesus monkey	5.5

Given the information in the table, it is reasonable to conclude that

A a comparison of the proteomes of humans and chimpanzees would show a 1.6% difference in amino acid sequences.

B gibbons and rhesus monkeys have more sequences in common than do humans and chimpanzees.

C humans and gibbons diverged more recently than did humans and rhesus monkeys.

D rhesus monkeys have more DNA sequences in common with chimpanzees than with humans.

Question 13 ⬤⬤⬤

The following table shows the number of amino acid differences in a protein molecule that is found in many species.

Species compared	Number of amino acid differences
Human vs chimpanzee	0
Human vs gorilla	2
Human vs monkey	12
Monkey vs chimpanzee	12
Monkey vs gorilla	14

On the basis of these results, it is reasonable to conclude that

A humans and gorillas may differ in more than two bases in the DNA sequence that codes for this protein.

B humans and chimpanzees have an identical DNA sequence in the region that codes for this protein.

C monkeys and gorillas share a more recent common ancestor than do humans and gorillas.

D gorillas are more closely related to chimpanzees than humans are to chimpanzees.

Short-answer questions

Question 14 (12 marks) ⬤⬤⬤

Fill in the diagram. Categorise each circle as either 'Mammals', 'Primates', 'Hominins' or 'Hominoids'. Then categorise the following traits by placing them inside the appropriate circles.

- Erect posture and bipedal locomotion
- The presence of hair or fur
- Prehensile hands and feet, usually with opposable thumbs and great toes
- Large brain
- Long arms relative to legs
- Glands specialised to produce milk, known as mammary glands
- No tail
- Flattened nails instead of claws on the digits

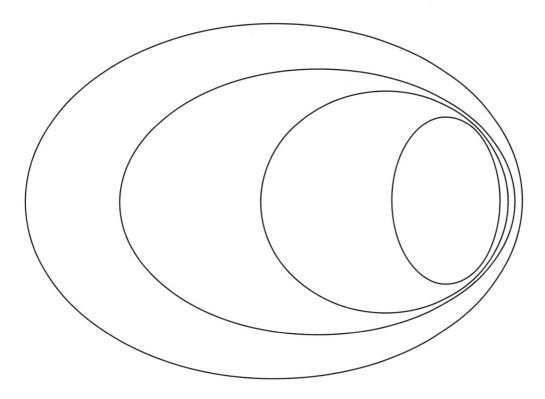

Question 15 (7 marks) ●●

The first primates were tree-dwelling insectivores who lived about 80 million years ago. In the following 50 million years, these primates evolved many adaptations for tree dwelling. One of the most important of these was forward-facing eyes and binocular vision. Between 20 and 30 million years ago, a group of primates, ancestral to humans and the great apes, evolved in Africa. Approximately 4 million years ago, the first hominins appeared.

a   What are **two** features, other than binocular vision, that are common to members of
     the primate order?   2 marks

b   **i** What are **two** features of early hominins that distinguish them from other primates?   2 marks

     ii For **one** of these features, outline how it would have been a selective advantage for
         early humans.   1 mark

     iii What is the species name given to Species C (found largely in Asia) in the table?   1 mark

Hominin skull	Brain size mean (cm^3)
Species A *Australopithecus afarensis*	450
Species B *Homo ergaster*	700
Species C	900

»

››	Hominin skull		Brain size mean (cm³)
	Species D *Homo sapiens*	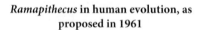	1300

c Describe the trend in brain size from *Australopithecus* species through to *Homo* species.   1 mark

Question 16 (8 marks)

Human evolution can be separated into two broad periods of time. The first is prior to the 1940s, when the recovery of the first non-modern human fossils occurred, resulting in the first attempts at reconstructing family trees of the transition from ape to human. The second is from the 1940s until the present day.

a Much more has been accomplished in reconstructing the phylogenetic tree of human evolution in the second period of time than the first. Why might this be the case?   1 mark

b Name **two** types of evidence used when trying to construct the evolution of modern humans?   2 marks

c Modern humans share 99% of their DNA sequences with chimpanzees and bonobos. What does this say about the evolutionary relationship between them?   1 mark

In 2010, a finger bone was discovered in a Siberian cave. Upon nuclear DNA analysis, it was found that the fossil was closely related to Neanderthals but could be classified as a new species of ancient humans. According to one theory, modern humans, Neanderthals and Denisovans all evolved from a homo species called *Homo heidelbergensis*. The theory states that *Homo heidelbergensis* migrated from Africa. One group travelled to north-west Asia and evolved to become the Neanderthals and another group travelled east to become the Denisovans. Modern humans descended from *Homo heidelbergensis* but evolved within Africa. Modern humans migrated from Africa at a much later time.

d Until 2010, the Denisovans were unknown to scientists. What does this tell you about the human fossil record?   1 mark

e Do you think the discovery of a new fossil could completely change the timeline of human evolution currently constructed? Explain.   2 marks

In 1932, in the Siwalik Hills in India, two parts of a fossilised upper jaw were discovered. The species to which they belonged was given the name *Ramapithecus*. Little attention was paid to the find at the time, but when the fossil was re-examined in 1961 it was concluded that they were those of an ancient hominin, showing more human-like than ape-like features. *Ramapithecus* was placed proudly on the human line of evolution.

Ramapithecus in human evolution, as proposed in 1961

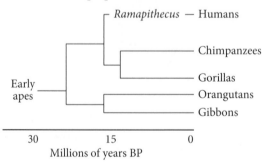

By the late 1970s, doubt was increasing over the conclusions reached about *Ramapithecus* in 1961. By 1982 it was widely accepted that *Ramapithecus* was an extinct relative of the orangutan, and not closely related to the hominins.

f Redraw the evolutionary tree above, to show the new position of *Ramapithecus*.   1 mark

Question 17 (8 marks) ©VCAA VCAA 2015 SB Q11 ●●●

Fossil evidence indicates that, between 30 000 and 80 000 years ago, populations of the two hominin species – modern humans (*Homo sapiens*) and the extinct Neanderthals (*Homo neanderthalensis*) – lived close to one another in parts of the Middle East, Europe and Asia. Researchers have constructed a theory about the relationships between ancient populations. This is represented in the following phylogenetic tree.

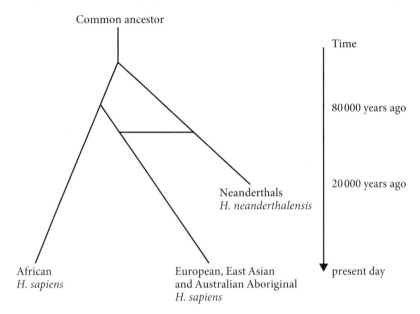

Recent DNA evidence has shown that:

- the genome of living humans of African descent does not contain Neanderthal DNA

- the genomes of living humans of European, East Asian and Australian Aboriginal descent all contain small amounts of Neanderthal DNA (1–4%).

a i Suggest how DNA from *H. neanderthalensis* entered the genome of present-day European, East Asian and Australian Aboriginal *H. sapiens*, and continues to be found in modern populations.   2 marks

ii What implication does this DNA evidence have for the classification of the two hominin species, *H. sapiens* and *H. neanderthalensis*, according to the common definition of a species?   1 mark

b There are several theories about the geographical origins of *H. sapiens*. Scientists consider that the absence of Neanderthal DNA in present-day African *H. sapiens* lends support to one theory about the geographical origins of *H. sapiens*. Name this theory and explain how the recent DNA evidence provided (above) supports it.   3 marks

c Consider the map below.

What does the DNA evidence provided suggest about the route and timing of the migration of the first Australian Aboriginals to arrive in Australia?   2 marks

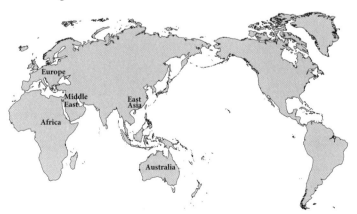

Question 18 (8 marks)   ©VCAA   VCAA 2008 (2) SB Q8 (adapted)   ●●▬

Two palaeoanthropologists each used fossil data to draw a model of the human evolutionary tree. The two models they produced are shown below.

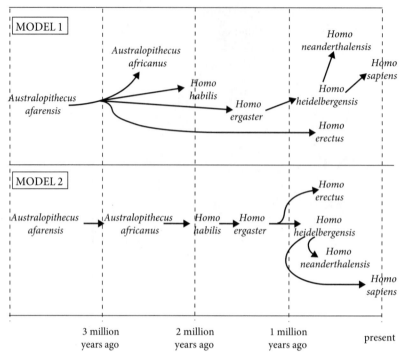

| 3 million years ago | 2 million years ago | 1 million years ago | present |

a   Explain how it is possible that the paleoanthropologists produced different models for the human evolutionary tree.   1 mark

b   **i**   State **one** feature of agreement between the models.   1 mark

    ii   State **one** feature of conflict between the models.   1 mark

c   Provide **two** structural features that would distinguish between the fossils of *Homo erectus* and *Australopithecus afarensis*.   2 marks

Scientists have recently discovered a tiny fossil skull in Indonesia. It has been named *Homo floresiensis* (the hobbit) and dated to the time our own ancestors were colonising the world. Some scientists believe *Homo floresiensis* evolved from *Homo erectus*. Fossils of *Homo erectus* have also been discovered in Indonesia.

d   Modify **one** of the models from the beginning of this question to include *Homo floresiensis*.   1 mark

    There is still some debate about what the hobbit is. Two explanations have been proposed.

    •   Explanation 1 – The hobbit belongs to a species of small-brained dwarf humans.

    •   Explanation 2 – The hobbit is a Stone Age *Homo sapiens* with a disease that stunts brain development.

e   **i**   Suggest **one** piece of evidence that would support Explanation 1.   1 mark

    ii   Suggest **one** piece of evidence that would support Explanation 2.   1 mark

Chapter 5 Area of Study 3
How is scientific inquiry used to investigate cellular processes and/or biological change?

Area of Study summary

This area of study is where you get the chance to apply all the practical skills you have learnt while studying VCE Biology. You will design or adapt a scientific investigation that relates to cellular processes and/or biological change and continuity over time, in either Unit 3 or Unit 4, or across both Units 3 and 4.

The investigation involves developing a question, writing an aim, making a hypothesis and planning a method that will allow you to answer the initial question. Your method must comply with safety and ethical guidelines. As you conduct your investigation, you will collect primary quantitative data, analyse and evaluate the data/results and consider the limitations of your results and the method used to obtain them. To reach a conclusion to your question, you will link your results to your learning in Unit 3 and/or Unit 4 and suggest further investigations that may help to provide further evidence to support your conclusion.

Presentation will be in a scientific poster format and you must keep your logbook of practical activities continually updated throughout the process. This is so your work can be authenticated and assessed.

Area of Study 3 Outcome 3

On completing this outcome, you should be able to:
- **understand** safety and ethical guidelines
- **create** questions and aims, and **formulate** hypotheses
- **apply** scientific skills to conduct investigations to collect and record data
- **analyse** and **evaluate** data, methods and biological models and communicate scientific ideas.

5.1 Scientific investigation

During your VCE Biology course, you will carry out an experiment designed or adapted by you and/or other members of your class. In order to complete this part of the course you will need to:
- develop a question and formulate a hypothesis
- plan an experiment or field investigation to answer the question
- carry out the experiment and collect data
- evaluate the data, including commenting on any possible limitations or errors
- present the data in a clear and logical way
- reach a conclusion that relates to your question and hypothesis
- present your findings in a scientific poster.

You will receive more information about this from your teacher and will have several hours to complete the experiment and poster. The task is assessed and forms part of your marks for Biology in the same way as any SAC. In addition to the assessment of your investigation and poster, this area of the course is examinable, so expect questions about hypotheses, experiment design, data and ethics in the end-of-year examination. The following sections summarise some of the key knowledge in the area of practical investigations. Much of it deals with definitions you need to know.

5.2 Scientific method

The scientific **method** refers to the steps taken to carry out an investigation. Sometimes in the Biology exam you may be asked to assess the method or design of an experiment, suggest an experiment to carry out, or suggest hypotheses to be tested. When designing experiments and/or hypotheses there are a few things to keep in mind.

5.2.1 Formulating a question

Choose one specific factor to investigate and frame your topic as a question.

Example of poor question	✓ Example of high-quality question
A poor question is general and includes many possibilities. For example: What makes a plant grow well?	A high-quality question is specific. For example: What effect does potassium availability (*one* variable) have on the mass (*one* thing to measure) of bean plants (*one* type of plant)?

Having chosen one factor to investigate, you make a hypothesis about the effect of varying that factor. For example, if you are investigating the effect of potassium on the growth of bean plants, you might make a simple statement: 'That the growth of bean plants improves with the addition of potassium to the soil'. Statements can be of the 'If ... then ...' type. '*If* increased potassium improves the growth of pea plants, *then* the more potassium that's added, the greater the growth of bean plants will be, as measured by their dry mass.' Another example: '*If* the enzyme works at an optimum rate in acidic environments, *then* the product will be produced more quickly in a solution at pH 4 than in a solution at pH 9.'

5.2.2 Variables

Any experiment should contain only *one* **variable** that changes; all other factors in the setup of your experiment should be constant. This is known as a **controlled experiment**. For example, in designing an experiment to investigate a factor affecting the rate of a biological reaction, you would look at just *one* factor. You may believe that the reaction rate will be affected by pH, temperature, substrate concentration, enzyme concentration or the shape of the container. You would choose *one* of these to investigate in an experiment.

The factor you choose to investigate is known as the **independent variable**. The **dependent variable** is the measurable or observable value that changes due to changes in the independent variable.

All other factors that have the potential to affect the outcome of the experiment need to be controlled. If you are testing a hypothesis about the effect of temperature on reaction rate, all other variables must be constant. Your samples will have different temperatures but they must have the same pH, the same light exposure, the same concentration of reactants, the same size and shape container, the same ... anything else you can think of! These are the **controlled variables**.

Some experiments require a control group or condition. Imagine that you are testing the hypothesis that the addition of potassium to soil improves the growth of plants. You get several plants and add potassium to the soil. They grow beautifully! You think you have supported your hypothesis, but what about a comparison with plants with no potassium added to the soil? This type of experiment requires a control. Another group of the same plants needs to be kept in identical conditions, but not be given potassium. *If* the plants with added potassium grow better than the control plants, *then* you have supported your hypothesis. A control is a group that is in the same conditions as the experimental group but does not get the treatment; that is, the independent variable is not applied to the control.

Finally, the results should be able to be repeated over a number of trials. This may involve using a number of samples at a time, by repeating the experiment several times, or by giving the experimental outline to others to test and verify.

If you do all the right things in your controlled experiment and the results are exactly as you predicted, then your hypothesis is *supported*; it is never proved. (After all, there may have been some unknown factor you overlooked.) If the results are contrary to your prediction, then the hypothesis is *disproved*.

5.3 Data quality

The quality of any data collected in an experiment is subject to many sources of potential error. Try to eliminate sources of error in your own experiment design, and be aware of them when you are evaluating data collected by others and their conclusions. The table below summarises things you need to consider when assessing the quality of experimental results.

Precision	A measurement is **precise** when it is carefully observed and recorded, using the available features of the instrument you are using. For example, if you are using a measuring tape marked in millimetres and you carefully read and record the result as 46 mm in several trials, you are being precise. If you say it's about 5 cm, you are being less precise. If you decide to add 10 grams of substrate to each of your experimental tubes and you make sure the balance reads exactly 10 grams, you are being precise.
	Being precise doesn't necessarily mean you are being accurate. If there is something wrong with your measuring tape or electronic balance, you may be precise, but not accurate.
	A caution about precision! You should not represent experimental data with values that exceed the precision of your measuring device. For example, say you measure the length of a number of worms using a standard ruler marked in millimetres. You use a calculator to work out the average length. The calculator says: 7.32456732. Don't include all the numbers after the decimal point. Round off to the value you could actually measure with your ruler. If you are using an electronic balance that measures to hundredths of a gram, any averages should be in hundredths of a gram, too.
Accuracy	A measurement is **accurate** when it is close to the actual true value. For your measurements to be accurate, not only does your measuring device need to be accurate, but you also need to use it properly. Some factors that may affect accuracy are: • a poorly calibrated measuring device • leaks in your system, especially when collecting gases • observation errors, such as parallax error (looking at a dial or marker from the wrong angle) • errors in the setup of the experiment, such as wrong quantities used, contaminated equipment or incorrectly labelled reactants. Reducing errors increases accuracy.
Repeatability	Repeatability refers to the claimed results of an experiment. Any significant results must be more than a one-off finding and must be **repeatable** by you using the same method and equipment, under the same conditions with identical test material.
Reproducibility	Reproducibility is very similar to repeatability; however, it refers to the similarity of independent results under different conditions. That is, the results must be **reproducible** by another operator, in a different laboratory.
Reliability	If your data is repeatable and reproducible, it is deemed to be **reliable**. Reliability refers to the consistency of the results of a study when repeated or reproduced.
Validity	Results and conclusions are considered scientifically **valid** only if all the steps in the scientific method are followed, controls are used, and your conclusion is supported by all the data. If results are valid, they measure what they are intended to measure.

››

>> **Bias**	**Bias** is intentionally or unintentionally influencing the outcome of an experiment by the choices you make as the experimenter. Bias can occur at many points in the experimental process. It can be a particular problem when choosing subjects for an experiment. Imagine you wanted to determine the average height of 17-year-old males. Rather than designing a process to select boys randomly, you just ask your mates to be subjects. This is bias and could influence your result (especially if they are mates from your basketball team). In another example, you want to test the health benefits of acai berries. You call for volunteers, and all your applicants are already fit, active, healthy young people. Alternatively, many people apply and you exclude the older, bigger people from the experiment. In a school experiment your bias might come out when you decide to discard a result that doesn't support your hypothesis, rather than examine the experiment design or repeat a process.

5.3.1 Sources of error

Even with the best methodology, the possibility of errors cannot be eliminated by repeating the experiment. There will always be some doubt associated with the results obtained from measurement. The two types of errors that should be considered for VCE Biology are random errors and systematic errors. Understanding how to reduce the errors will improve the accuracy and precision of your results.

Random error	**Random errors** affect the precision of results. They occur in measurements and can be minimised by using correctly scaled laboratory equipment and by repeated trials and calculating the mean. For example, using a graduated cylinder is a more precise tool than a beaker for measuring fluids because it has more subdivisions between the millilitre marks.
Systematic error	**Systematic errors** affect the accuracy of results. They occur when using equipment to gain a reading and the equipment provides a result that is shifted from the true value. They cannot be improved by repeated readings. Inaccurate readings can result from not calibrating equipment. For example, forgetting to tare or zero scales will produce results that are always off by the same mass.
Parallax error	Incorrect reading of results, also known as **parallax error**, can be a random or systematic error. A parallax error is where the reading of a result can change depending on the angle from which you view it. This is why you should always view the meniscus at eye level. If the angle from which you view the result is always the same, but is incorrect, it is a systematic error. If the angle from which you view the result is different each time, and is incorrect, it is a random error. 7.0 mL 6.6 mL 6.2 mL

It should be noted that **personal errors**, such as mixing up readings, mistakes or miscalculations can be eliminated by repeating the experiment again correctly. Personal errors do not form part of the analysis of uncertainties. If you have any anomalous results or outliers in your data (that is, any unusual values in the data set), it is possible that there is a personal error. Rather than discounting the result, extra readings may help to identify the source of error.

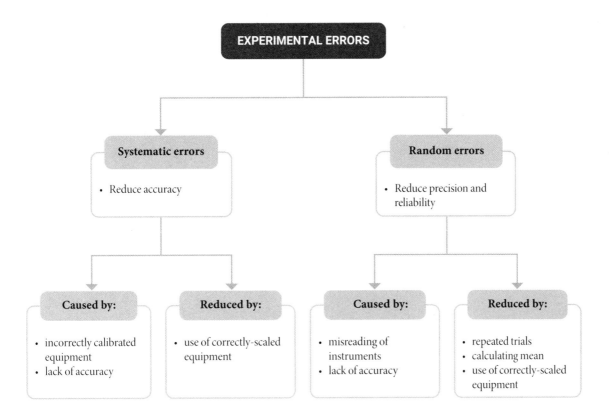

5.4 Data presentation

Results of investigations are often presented graphically, in tables, charts and graphs. In the Biology exam, you will often be asked to analyse graphs and interpret information presented graphically – and this is often poorly done by students. Similarly, when you design your scientific poster you will need to include graphical presentations of your own data.

5.4.1 Reading graphs

The most important thing to consider when examining data in any format is to read all the information in the graph or table, particularly labels and scales. The graph below highlights the important points.

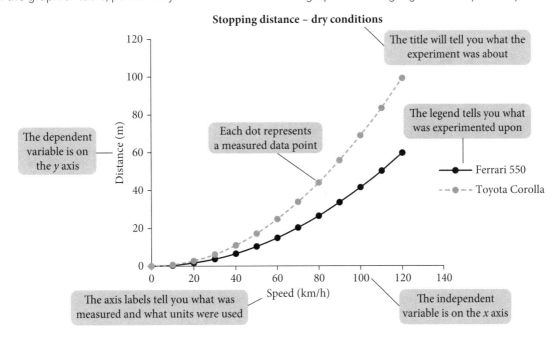

Once you have reviewed all the information, you are ready to comment on the data. Focus on what the graph actually shows, rather than conclusions it may lead you to make. For the above example, you would not say the graph shows that Ferrari is better than Toyota! You need to be specific. For example:

- The graph shows that, at all speeds tested, the Ferrari was able to stop in a shorter distance than the Toyota.

- The difference in stopping distance increased as the speed increased.

- At 100 km/h, the Ferrari stopped at 41 metres, whereas it took the Toyota 70 metres to stop.

5.4.2 Data presentation formats

When designing your poster for your own investigation, don't overdo data presentation. Scientists are unimpressed with multiple presentations of the same data. One of the skills you'll be assessed on is your ability to choose the best format in which to display your data. The following table has some examples of when to use particular graph types.

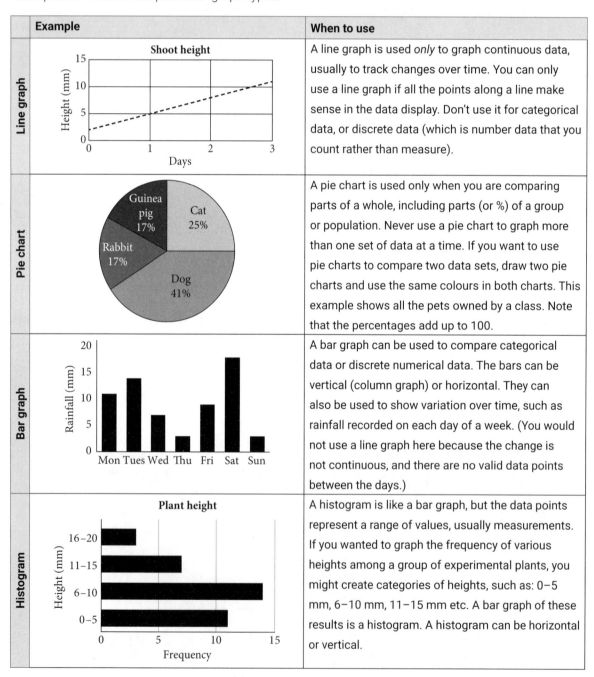

	Example	When to use
Line graph	Shoot height	A line graph is used *only* to graph continuous data, usually to track changes over time. You can only use a line graph if all the points along a line make sense in the data display. Don't use it for categorical data, or discrete data (which is number data that you count rather than measure).
Pie chart	Guinea pig 17%, Cat 25%, Rabbit 17%, Dog 41%	A pie chart is used only when you are comparing parts of a whole, including parts (or %) of a group or population. Never use a pie chart to graph more than one set of data at a time. If you want to use pie charts to compare two data sets, draw two pie charts and use the same colours in both charts. This example shows all the pets owned by a class. Note that the percentages add up to 100.
Bar graph	Rainfall (Mon–Sun)	A bar graph can be used to compare categorical data or discrete numerical data. The bars can be vertical (column graph) or horizontal. They can also be used to show variation over time, such as rainfall recorded on each day of a week. (You would not use a line graph here because the change is not continuous, and there are no valid data points between the days.)
Histogram	Plant height	A histogram is like a bar graph, but the data points represent a range of values, usually measurements. If you wanted to graph the frequency of various heights among a group of experimental plants, you might create categories of heights, such as: 0–5 mm, 6–10 mm, 11–15 mm etc. A bar graph of these results is a histogram. A histogram can be horizontal or vertical.

Example	When to use
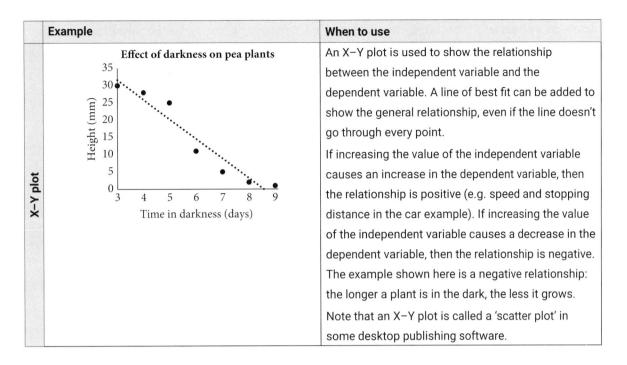	An X–Y plot is used to show the relationship between the independent variable and the dependent variable. A line of best fit can be added to show the general relationship, even if the line doesn't go through every point.
	If increasing the value of the independent variable causes an increase in the dependent variable, then the relationship is positive (e.g. speed and stopping distance in the car example). If increasing the value of the independent variable causes a decrease in the dependent variable, then the relationship is negative. The example shown here is a negative relationship: the longer a plant is in the dark, the less it grows.
	Note that an X–Y plot is called a 'scatter plot' in some desktop publishing software.

(Left margin label: X–Y plot)

5.5 Experimenting ethically

All experiments and scientific investigations need to be carried out ethically. In Victoria, any biomedical research must be evaluated by an ethical review before it can be carried out. This is a formal process that assesses whether research proposals meet ethical standards. Review committees from our major hospitals generally carry out this work.

In science, **ethics** is a broad term referring to the expected conduct of the experimenter in designing and implementing the experiment and in publishing its results. If you are studying Psychology you will have spent a lot of time looking at ethics, as psychological experiments often involve human subjects and sensitive issues. The points below summarise some of the key ethical considerations in experimental design and practice.

Experimental practice	Be competent and careful. Make sure you know what you are doing; do the background research; keep up to date and do your best to be careful and accurate.
Integrity	Be honest, objective and open. Report results as you found them. Do not include false data or omit data because it doesn't suit your objectives. Share your methods and results openly.
Respect for others, especially human subjects	Your experimental design should minimise risks or harm. • Avoid discrimination based on sex, ethnicity, disability or age (except where essential for the research). • Subjects need to be informed of the purpose and design of the experiment and should give informed consent. (This does not apply to anonymous data such as an investigation of the number of positive results in infant genetic screening.) • Take extra precautions with vulnerable groups. • Protect the confidentiality and/or anonymity of subjects and the security of sensitive information. • Allow subjects to leave the experiment at any time.
Research merit	Weigh up whether the potential outcome (new knowledge) is worth the potential risk.
Health and safety	You should ensure the safety of anyone involved in your experiment, including yourself. Use safe practices and wear appropriate safety clothing.

(Right margin label: U4 – AREA OF STUDY 3)

››	Intellectual property	Don't copy other people's work. Give proper acknowledgement to the ideas or work of others that have helped you in your work.
	Use of animals	Scientists, teachers and students are subject to the *Australian code for the care and use of animals for scientific purposes, 8th edition 2013*. It sets out principles for the care and use of animals. The main principles are as follows. • Avoid or minimise harm, including pain and distress, to animals. • Support the wellbeing of animals. • Use animals only when justified. • Apply high standards of scientific integrity. • Apply Replacement, Reduction and Refinement (the three Rs) at all stages. Think of a *replacement* for the use of animals in your experimental design. If the use of animals is unavoidable, try to *reduce* the number of animals you use. *Refine* your procedures and techniques to minimise any harm to animals.

5.6 Science communication – your poster

The poster you submit for your own investigation must include all of the following features.

> **Hint**
> You should avoid using personal pronouns in a scientific report. This helps remove any subjectivity or opinion and keeps your findings objective to the data collected.

Heading	This is your research question. Ensure it finishes with a question mark (?). A useful format to follow: How does the (insert IV) effect the (insert DV) if/when … ?	Write in present tense. The heading should be one sentence in length.
Introduction	This must include a clear aim and hypothesis. Your hypothesis should be a testable statement. A useful format to follow for this is: If the (insert DV) is affected by the (IV), then (insert the effect on the DV) when (insert the change in IV). For example: If the enzyme activity is affected by the temperature, then the enzyme activity will decrease when the temperature is increased above 60°C. You should also include a reason for doing the experiment, and how it may contribute to scientific knowledge and application, linking this to scientific concepts covered in Unit 3 and/or Unit 4.	Write in past tense.
Methodology	This is a step-by-step account of how you conducted the experiment. Be sure to include all equipment, including how you calibrated it and all measurements/quantities, with units.	Write in past tense. You can include a labelled diagram of your experimental set-up.
Results	Here you want to clearly communicate your data, showing trends or patterns. To do this, you must transform your raw numbers and present them visually using a graph. Under the graph you describe the relationship between the IV and the DV but do not explain it … yet!	Write in past tense.
Communication statement	Report the key finding of the investigation as a one-sentence summary that answers the question under investigation.	Write in past tense.

Discussion (analysis or results)	This section is for explaining your results and interpreting the data. You could do this by stating your results, and whether the data supports your hypothesis or not. You can then explain why, incorporating strengths and weaknesses of your experiment. Discuss errors, including accuracy, precision and any limitations. You should suggest any improvements.	Write in present tense.
Conclusion	A conclusion should restate whether your hypothesis was supported or not, any limitations or improvements for next time and include any implications for your findings on scientific knowledge, society, and/or the environment, linking this to the relevant Biology studied in Unit 3 and/or Unit 4.	Write in present tense.
References and acknowledgements	This is a formatted list of quotations and sourced content in alphabetical order.	N/A

> **Hint**
> If your hypothesis is not supported – that is okay! Just explain why.

Before you submit your poster, check the following to ensure your scientific communication is of the highest standard.

- When presenting your research, be sure to communicate your findings logically and clearly, with each section such as 'Aim' and 'Hypothesis' easily identified with headings and subheadings.

- Ensure your information is concise and relevant to the investigation and your findings. Remain objective to the results and avoid personal pronouns such as 'I', 'we', 'my', 'our' and 'me'.

- Label all images and graphs clearly using relevant scales for graphs, annotations and labels. Labels should be in sentence case with correct units included. If you have a table of results, ensure there are no units in the body of the table (only in the heading) and that all values are consistent to the same number of decimal places.

- Avoid using the word 'amount'. Instead, choose more specific terminology such as 'volume', 'mass' or 'concentration', and ensure you use the correct units and symbols. Below are the seven common SI units, their standard abbreviations and units of measurement.

Property	Abbreviation/symbol	Units
Mass	kg	Kilogram
Time	s	Second
Volume	L	Litre
Temperature/thermal energy	K	Kelvin
Length	m	Metre
Concentration	mol	Mole
Luminous intensity	cd	Candela

- Ensure all references are included and formatted correctly. To reference a journal article:

  1 Author or authors; surname first, followed by the first initial

  2 Year of publication of the article

  3 Title of article in single inverted commas

  4 Journal title in italics

  5 Volume of journal

  6 Issue number of journal

  7 Page range of article

  For example: Martin, S. 2020, 'How is scientific inquiry used to investigate cellular processes and/or biological change?', *A+ Biology Study Notes*, 1e, pp. 100–15.

- Finally, proofread your work prior to submission!

Glossary

accurate Describes a measurement that is as close to the true value as possible

bias An error that occurs when an investigation is not randomised, particularly if the investigator is affected by their own expectations of the outcome

controlled experiment An experiment where only one variable is manipulated

controlled variable A variable that is kept constant during an investigation in order to determine the relationship between the independent and dependent variables

dependent variable The variable that is measured and whose value depends on the independent variable; i.e. it responds to the independent variable

ethics A system of moral principles that considers what is good and bad for society

independent variable The variable that is changed or manipulated by the scientist and assumed to have an effect on the dependent variable

method The steps taken to carry out a scientific investigation

parallax error The difference in the apparent position of something when viewed from different angles

personal error A mistake or miscalculation due to human error

precise Describes a measurement observed and recorded carefully; when a process, procedure or step

A+ DIGITAL FLASHCARDS
Revise this topic's key terms and concepts by scanning the QR code or typing the URL into your browser.

https://get.ga/a-biology-vce-u34

is repeated the same way each time to ensure the same (or very similar) results each time

random error An unpredictable variation in measurement; can be improved by taking multiple measurements and calculating an average

reliable Describes an investigation or data that is reproducible and repeatable

repeatable Describes an investigation that can be conducted again by the same investigator under the same conditions to generate similar results

reproducible Describes an investigation that can be conducted by a different investigator under different conditions to generate similar results

systematic error A predictable deviation in data; e.g. as a result of the equipment used

valid Describes results that are affected by only a single independent variable and hence measure what they are intended to measure

variable Something that can change or be changed, as distinct from a constant, which does not change

Exam practice

Solutions start on page 253.

Multiple-choice questions

Question 1

Precision is

A how close measurements of the same item are to each other.

B how close a measurement is to the true or accepted value.

C the closeness of agreement between independent test results, obtained with the same method, on the same test material, in the same laboratory, by the same operator, and using the same equipment within short intervals of time.

D how well a scientific test or piece of research actually measures what it sets out to, or how well it reflects the reality it claims to represent.

Question 2 ▣□□

Reproducibility is

A how close measurements of the same item are to each other.

B how close a measurement is to the true or accepted value.

C the closeness of agreement between independent test results, obtained with the same method, on the same test material, in the same laboratory, by the same operator, and using the same equipment within short intervals of time.

D the extent to which consistent results are obtained when an experiment is repeated.

Use the following information to answer Questions 3–6.

Elodea is an aquatic plant, often called waterweed, that is native to the Americas and is often used in aquariums. As *Elodea* photosynthesises, carbon dioxide and oxygen are absorbed from the surrounding water. During cellular respiration, carbon dioxide is released, where it reacts with water to create carbonic acid and therefore decreases the pH of the water surrounding the plants.

Bromothymol blue can be used to monitor photosynthesis and cellular respiration. This indicator changes colour around pH 7. Higher than pH 7, the indicator turns blue, and below pH 7 (when the solution becomes more acidic), the indicator turns yellow.

A student wanted to investigate photosynthesis and cellular respiration. She set up three test tubes, each with with a 10 cm long piece of *Elodea* and 15 mL of water. Bromothymol blue was added to each test tube. It immediately turned blue/green in colour. The student immediately bubbled air through the water with a straw until the water turned yellow. The student placed one test tube in the window so it could experience the day/night cycle, one tube in compete darkness and the other under a permanent light. The test tubes were set up in the first lesson of the school day at 9 a.m. Results were gathered at 3.30 p.m. and then again at 9 a.m. the following morning.

The results are below.

Condition	Initial colour	Colour at 3.30 p.m.	Colour at 9 a.m.
Day/night cycle	Yellow	Blue	Green
24 hour darkness	Yellow	Yellow	Yellow
24 hour light	Yellow	Blue	Blue

Question 3 ▣▣□

An explanation for the results of the experiment would be that

A the rate of photosynthesis is higher in the tube experiencing the day/night cycle.

B the tube experiencing complete daylight will continue to photosynthesise for the entire 24-hour period, increasing the levels of carbon dioxide.

C the tube experiencing total daylight would have a higher rate of cellular respiration than photosynthesis.

D all tubes have an equal rate of photosynthesis.

Question 4 ▣▣□

When experiencing the day/night cycle, at 9 a.m. the following day, the tube turned green. This is because

A the chlorophyll from the *Elodea* plant leached out into the water to turn it green overnight.

B more carbon dioxide was produced overnight, changing the colour of the indicator to green.

C the rate of respiration is slower at night, resulting in the pH of the water increasing.

D the rate of respiration is slower than the rate of photosynthesis at night, using up carbon dioxide in the water.

Question 5 ◑▨▨

An improvement that could be made to this experiment could be to

A increase the size of the piece of *Elodea*.

B blow more carbon dioxide into the water.

C increase the number of trials under each condition.

D decrease the volume of water used in the experiment.

Question 6 ◐◐▨

Which of the following statements is correct?

A The experiment is precise because it measured the pH of the water.

B The experiment accurate because the results are close to the true values.

C The results are valid because it measures what was intended.

D The results are reproducible because the experimental design was quite simple.

Use the following information to answer Questions 7 and 8.

Four groups of students carried out an experiment in which the effect of glucose concentration on the fermentation rate of yeast was measured. The fermentation rate was determined by the rate of temperature change of the fermenting mixture. Before beginning the experiment, each group practised measuring the temperature of water and checked the group's thermometer against an electronic thermometer that gave a true measure of temperature. The following results were obtained during the practice.

Group	Each group's thermometer readings (°C)			Electronic thermometer reading (°C)
	1st measurement	2nd measurement	3rd measurement	
1	18.0	17.0	17.5	20.1
2	18.0	18.0	18.5	20.5
3	21.0	21.0	20.5	19.9
4	18.0	19.0	21.0	20.2

Question 7 ©VCAA VCAA 2018 SA Q11 ◐◐◐

Which one of the following statements is correct?

A Group 1's measurements are the most accurate but the least precise.

B Group 2's measurements are accurate but not precise.

C Group 3's measurements are precise but not accurate.

D Group 4's measurements are both accurate and precise.

Question 8 ©VCAA VCAA 2018 SA Q12 ●●●

Each group conducted the experiment three times (Trial 1, Trial 2, Trial 3). Five different concentrations of glucose were used in each trial. Each group plotted its results on a graph. The black bar represents Trial 1, the white bar represents Trial 2, and the striped bar represents Trial 3.

Which one of the following statements about the experiment's results can be concluded from the graphs?

A Group 1's results are more valid than the other groups', but less reliable.

B Group 2's results are less reliable, but more precise and accurate.

C Group 3's results are the most accurate and reliable.

D Group 4's results are more reliable than the other groups'.

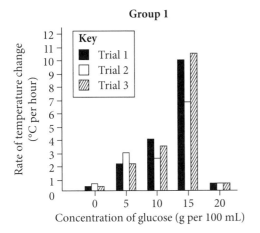

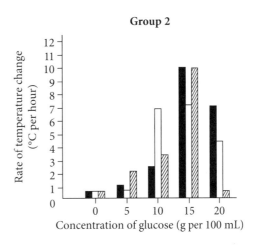

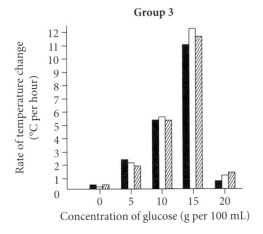

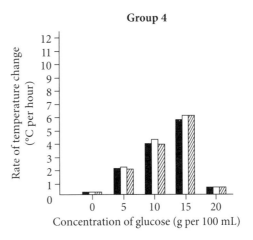

Question 9 ●○○

The independent variable is the variable that

A the experimenter changes within the experiment.

B is measured during the experiment.

C is independent of the experiment and can't be controlled by the researcher.

D needs to be controlled in the experiment.

Question 10 ⬤◦◦

A control is used in an experiment so that

A the change in the dependent variable can be attributed to the change in the independent variable.

B the results in the experiment can be compared to the results of the control.

C it can be used as a baseline for the experiment.

D All of the above

Question 11 ⬤⬤◦

A scientist wanted to identify the mode of transmission of a particular human bacterial pathogen. How might he design an experiment to help identify this?

A Expose a number of people to the pathogen in a number of different ways to see who becomes infected.

B Place some bacteria in water and then spread a measured amount of water on an agar plate to see if the bacterium is water borne.

C Expose rabbits to the pathogen in a variety of ways to see which one becomes infected.

D Identify the pathogen's gene sequence using PCR.

Question 12 ⬤◦◦

An example of qualitative data is the

A concentration of carbon dioxide in an experiment investigating photosynthesis.

B volume of gas produced in an experiment investigating cellular respiration.

C presence or absence of antibodies for the SARS-CoV-2 virus.

D volume of biofuel produced through fermentation.

Use the following information to answer Questions 13–15.

An experiment was carried out by students to test the effect of temperature on the growth of bacteria. Bacterial cells were spread onto plates of nutrient agar that were then kept at three different temperatures: –10°C, 15°C and 25°C. All other variables were kept constant. The experiment was carried out over four days. The nutrient agar was observed every day at the same time and the percentage of nutrient agar covered by bacteria was recorded. At the conclusion of the experiment, the results were recorded in a table, which is shown below.

Time (days)	Percentage of nutrient agar covered by bacteria at three different temperatures		
	–10°C	**15°C**	**25°C**
0	0	0	0
1	0	5	10
2	0	10	20
3	0	15	40
4	0	20	60

Question 13 ©VCAA VCAA 2019 SA Q7 ⬤◦◦

Which one of the following hypotheses is supported by the results?

A If the bacteria grow for four days, then the nutrient agar will be completely covered in bacteria.

B If the bacteria are kept in the dark, then the bacteria will grow more slowly.

C If the bacteria grow faster, then the temperature of the location will increase.

D If the temperature increases, then the bacteria will grow more quickly.

Question 14 ©VCAA VCAA 2019 SA Q8 ●●●

In this experiment, the dependent variable is

A time.

B temperature.

C the number of bacterial cells.

D the percentage of nutrient agar covered by bacteria.

Question 15 ©VCAA VCAA 2019 SA Q9 ●●●

The students wanted to check the reliability of their data. The students should

A repeat the experiment several times to find out if they would obtain the same data.

B organise their data into a different format to help identify a trend.

C change the independent variable in the experiment.

D rewrite the method for completing the experiment.

Short-answer questions

Question 16 (15 marks) ●●●

Pepsin is one of the main digestive enzymes in the digestive systems of humans and many other animals. Six test tubes were set up to investigate the digestion of protein by pepsin. The contents of each test tube are listed below.

A: 1.0% pepsin in water

B: 1.0% pepsin in 0.4% hydrochloric acid

C: 0.4% hydrochloric acid

D: 1.0% pepsin in 0.5% sodium bicarbonate

E: 0.5% sodium bicarbonate

F: distilled water

A cube of egg white was added to each tube. Egg white is rich in protein. The tubes were allowed to sit for 12 hours. The results are shown below.

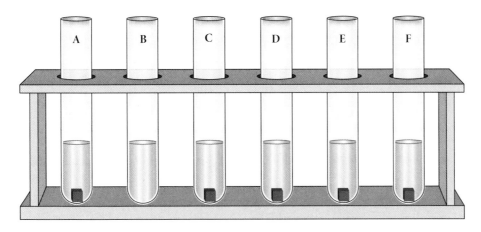

a Why was tube F included in this experiment? — 1 mark

b List **three** variables that need to remain constant in this experiment. — 3 marks

c A student hypothesised that 'in the presence of pepsin, all protein will be digested into smaller soluble peptides'. Do the results support this hypothesis? Explain. — 2 marks

d Describe the results of this experiment. — 3 marks

   e Propose **three** improvements to this experiment.   3 marks

   f Write a detailed method to include the improvements you identified in your answer
      to part **e**.   3 marks

Question 17 (13 marks) ●●○

In 1981, Barry Marshall began working with Robin Warren at the Royal Perth Hospital. Warren had discovered that the gut could be overrun by hardy, spiral shaped bacteria called *Helicobacter pylori*, which he proposed caused stomach ulcers and could easily be treated with antibiotics. However, mainstream gastroenterologists were dismissive of this hypothesis, holding on to the old idea that ulcers were caused by stress.

Unable to make his case in studies with lab mice because *H. pylori* affects only primates, Marshall ran an experiment on the only human patient he could ethically recruit: himself. He took some *H. pylori* from the gut of an ailing patient, stirred it into a broth, and drank it. Marshall developed gastritis, the precursor to an ulcer. After taking a biopsy from his own gut he cultured *H. pylori* from the affected tissue.

   a What is the hypothesis for this experiment?   1 mark

   b Comment on the validity and reproducibility of this experiment.   2 marks

   c Comment on the health, safety and ethical issues surrounding this experiment.   2 marks

To test this hypothesis and the discovery made by Marshall, scientists took biopsies from patients with duodenal ulcers. They found that 95% of participants were infected with *H. pylori*. In a second study conducted in Europe, 97% of participants with duodenal ulcers were also infected with *H. pylori*.

   d What are the dependent and independent variables in the investigations above?   2 marks

   e Comment on the results of these experiments.   2 marks

Further studies showed that, once treated with antibiotics, the risk of recurrence of ulcer disease in all participants reduced to less than 10%.

   f Write a conclusion to these investigations. Make sure to reference your hypothesis.   2 marks

   g Would it be beneficial and cost effective to create a vaccine for *H. pylori*? Explain.   2 marks

Question 18 (14 marks) ©VCAA VCAA 2017 SB Q11 ●●●

Matthew investigated how changes in environmental temperature affected oxygen (O_2) and carbon dioxide (CO_2) levels in the air around a cockroach. He used three digital probes linked to a computer, a closed animal chamber and a heat lamp in the experimental set-up shown.

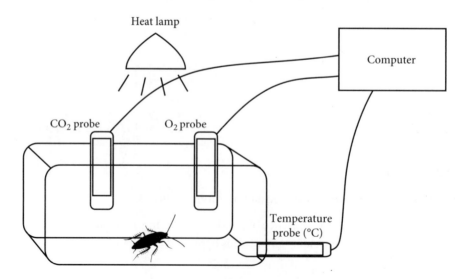

   a Name the cellular process being investigated in Matthew's experiment.   1 mark

b Identify the

   i dependent variables   1 mark

   ii independent variable.   1 mark

Before placing the cockroach in the chamber, Matthew decided to measure the temperature, and carbon dioxide and oxygen levels for four minutes. The following results were recorded.

Time (minutes)	CO_2 (%)	O_2 (%)	Temperature (°C)
0	0.04	22.3	29.5
1	0.04	22.1	29.8
2	0.04	22.0	30.0
3	0.04	22.0	30.0
4	0.04	22.0	30.0

c Explain why Matthew recorded the data for four minutes and not just one minute.   1 mark

After the initial four-minute period, Matthew quickly placed the cockroach in the chamber and began recording the data from the digital probes. After 10 minutes, he placed ice packs around the sides of the animal chamber to slowly bring the temperature of the chamber down to 10°C. He recorded the data using the digital probes for a further 20 minutes. He repeated the experiment once every day for the next six days with the same cockroach. At all times, he took care to ensure that the cockroach showed no signs of stress.

d Other than repeating the entire experiment, identify **two** control measures Matthew should have included in his experimental design. Explain how each of these control measures could affect the results if not kept constant.   4 marks

Matthew constructed the following graphs from the averaged results of the seven experiments.

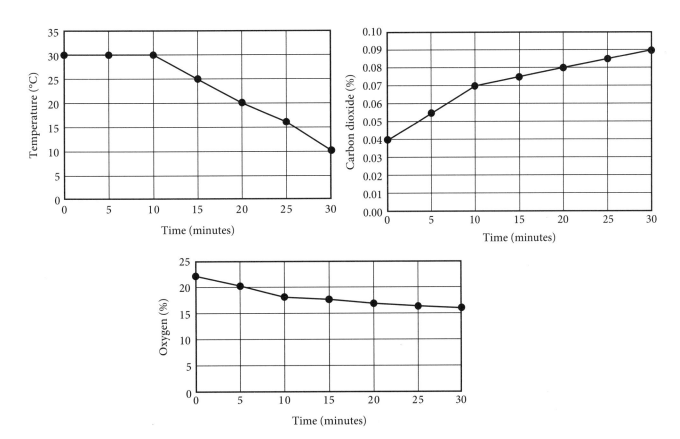

e **i** Using the graphical data, describe the changes in the levels of carbon dioxide and oxygen when the temperature in the chamber was kept constant compared to when the temperature was decreasing.   2 marks

ii What conclusion do you think Matthew can draw from his investigation? You should refer to each of the following in your response.

- the cellular process named in part **a**

- the variables identified in part **b**

- the evidence collected during Matthew's experiments   4 marks

Question 19 (19 marks) ⬤⬤⬤

Seed size in beans shows variation in natural populations and this variation is influenced by many genes. It has been found that there is a link between the size of the bean plant and the size of the seed it was grown from. A breeding experiment was carried out in order to determine how large beans could be grown. Initially three breeding populations of bean plants were established.

Group A: 20 plants bearing large beans

Group B: 20 plants bearing small beans

Group C: 20 plants chosen at random

Each population was bred under the same environmental conditions. Pollination was carried out artificially so that plants bred only within their own group. The plants were bred for 10 generations. For Group A, only the 20 plants bearing the largest beans were retained for breeding in the next generation; for Group B, only the 20 smallest were retained; and in Group C, 20 plants were chosen at random. The results of this breeding experiment are summarised in the table below.

Generation no.	Group A Mean seed mass (g)	Group B Mean seed mass (g)	Group C Mean seed mass (g)
1	2.6	1.8	2.3
2	2.8	1.7	2.4
3	3.0	1.6	2.2
4	3.1	1.4	2.2
5	3.3	1.3	2.1
6	3.5	1.2	2.3
7	3.8	1.1	2.2
8	4.0	1.0	2.4
9	4.0	1.0	2.3
10	4.0	1.0	2.3

a Write a hypothesis for this experiment.   1 mark

b Draw a graph showing the trend in the data above.   3 marks

c Describe the results in your graph.   3 marks

d What is the purpose of the Group C plants in this experiment?   1 mark

e List **three** variables that would need to remain constant throughout the experiment for all three groups.   3 marks

f Why were the environmental conditions kept the same for all three populations?   3 marks

g i How would you account for the fact that seeds from the Group A plants became larger in generations 2–8?   1 mark

ii Give a genetic explanation for the fact that the mean mass of the seeds stopped increasing after the eighth generation.   2 marks

h What name is given to this type of breeding program?   1 mark

i Outline **one** possible negative consequence of the application of this type of breeding program.   1 mark

Question 20 (9 marks) ©VCAA [VCAA Sample exam SB Q11] ●●●

Laura wanted to investigate the effectiveness of an antibiotic against the bacterium *Escherichia coli*. She prepared five different concentrations of the antibiotic. She wrote the following method.

1 Collect five agar plates containing nutrient agar.

2 Label each agar plate with the five different concentrations of the antibiotic.

3 Collect a sample of *E. coli* in a broth culture.

4 Put on a pair of disposable gloves.

5 Measure 0.5 mL of broth in a pipette and place in the centre of the first agar plate.

6 Spread the bacteria evenly over the agar plate with the spreader provided.

7 Place a drop of the antibiotic in the centre of the agar plate.

8 Close the lid of the agar plate and tape the lid to the bottom of the agar plate with sticky tape.

9 Repeat steps 6 to 8 with the other four concentrations of the antibiotic.

10 Place the agar plates on the side bench and leave overnight.

11 Wash your hands and dispose of the gloves.

a What hypothesis is Laura testing with this experiment?   1 mark

b Name the independent variable in this experiment. Justify your answer.   2 marks

c Identify **two** sources of experimental errors in this investigation and suggest how the experimental design could be changed to reduce the effect of these errors.   4 marks

d Laura wanted to repeat the experiment to test the effectiveness of an antiviral drug against *E. coli*. She prepared five different concentrations of the antiviral drug and followed the same steps that she used for the antibiotic. Explain the results that Laura would be expected to obtain.   2 marks

SOLUTIONS

CHAPTER 1, UNIT 3 AREA OF STUDY 1

The relationship between nucleic acids and proteins

Multiple-choice questions

1 A

Due to complementary base pairing, cytosine pairs with guanine; therefore, there will be the same number of each.

B is incorrect because whether there is twice the amount of cytosine as thymine depends on the sequence of bases on the segment of DNA. **C** is incorrect because it depends on the sequence of bases on the segment of DNA as to whether all four are in equal amounts. **D** is incorrect because there is no uracil in DNA.

2 D

A nucleotide consists of three subunits: deoxyribose sugar, phosphate and a nitrogenous base.

A is incorrect because DNA contains deoxyribose sugar, not ribose (this is in RNA). **B** is incorrect because DNA contains thymine, not uracil (this is in RNA). **C** is incorrect because DNA contains deoxyribose sugar, not ribose, and thymine, not uracil.

3 A

The opposite 5´ to 3´ orientations of the nucleotides in the 2D image reflects the anti-parallel arrangement.

B is incorrect because replication is not depicted. **C** is incorrect because DNA has deoxyribose (not ribose). **D** is incorrect because the 3D double helix cannot be seen.

4 D

Codons are present on the transcribed mRNA. Each amino acid may have multiple codons that code for the same amino acid.

A is incorrect because DNA has thymine (not uracil). **B** is incorrect because RNA has uracil (not thymine). **C** is incorrect because amino acids may have multiple codons as the code is degenerate, but a single codon only codes for one amino acid.

5 C

After messenger RNA has been transcribed, RNA processing occurs, which includes the removal of introns.

A is incorrect because modifications to the mRNA include the removal of introns, and the retention of different exons can be changed depending on the protein being produced. **B** is incorrect because the poly-A tail is added to the 3´ end of the RNA sequence. **D** is incorrect because introns are always removed.

6 C

The anticodon on the tRNA (UCC) is complementary to the mRNA sequence (AGG), which in turn is complementary to the template DNA strand (TCC).

A is incorrect because this is the mRNA sequence. **B** is incorrect because the mRNA codon is complementary to the anticodon and contains uracil. **D** is incorrect because thymine (T) would be complementary to adenine (A) on the mRNA strand.

7 A

There needs to be a way to regulate the transcription of genes, so the promotor and operator must come before the genes along the strand of DNA, as this is where regulatory proteins can bind.

B and **C** are incorrect because the genes come after the promotor and operator. **D** is incorrect because the promotor is first in the sequence, followed by the operator.

8 B

The promotor region is needed as a binding site on the DNA for the enzyme RNA polymerase so that the gene can be transcribed.

A is incorrect because the creation of an RNA copy from a DNA template is called transcription. **C** is incorrect because the promotor region is used in the process of transcription, not during splicing or identification of exons/introns. **D** is incorrect because a repressor protein will bind to the operator region.

9 C

The *trp* operon is regulated by a repressor protein that binds to tryptophan. Once this occurs, the conformation of the repressor protein changes, enabling it to bind to the operator region of the *trp* operon and turn the operon 'off'. In the absence of tryptophan, the repressor protein cannot bind to the DNA strand and therefore the operon is switched 'on'.

A is incorrect because, in the absence of tryptophan, the repressor protein cannot bind to the DNA strand and therefore the operon is switched 'on'. **B** is incorrect because this system has no effect on cell mass. **D** is incorrect because the *trp* operon is affected by the presence or absence of tryptophan in order to conserve energy and increase cellular efficiency.

10 B

The diagram shows the production of tryptophan. This results in the DNA being transcribed to produce the enzymes required to make tryptophan.

A is incorrect because the diagram shows the production of tryptophan, not repressor proteins. **C** is incorrect because RNA polymerase was used (not DNA polymerase). **D** is incorrect because transcription comes before translation in the process.

11 B

Attenuation occurs because transcription and translation occur together. The ribosome arrives at the STOP codon on the mRNA, preventing sections 1 & 2 from forming any hairpin loops but allowing sections 3 & 4 to form a hairpin loop creating the pulling force to wrench the mRNA away from the DNA.

A is incorrect because the repressor protein is not involved in attenuation.

C is incorrect because sections 1 & 2 are prevented from forming a hairpin loop. **D** is incorrect as adenine and uracil are bonded by double hydrogen bonds.

12 C

Primary structure involves the sequence of amino acids joined by peptide bonds.

A is incorrect because primary structure is formed by covalent (peptide) bonding. **B** is incorrect because secondary structure involves alpha helices and beta-pleated sheets formed by hydrogen bonding. **D** is incorrect because tertiary structure involves the properties of the R groups and a variety of different types of bonds as a result.

13 D

The joining of amino acids and glucose monomers are both endergonic (require energy).

A is incorrect because hydrolysis reactions are the separation/breaking of larger molecules into monomers (individual, smaller molecules). **B** is incorrect because these reactions absorb energy when the bonds are formed. **C** is incorrect because DNA ligase joins nucleotides in DNA (not amino acids or glucose).

14 C

Tertiary structure involves the properties of the R groups creating a variety of different bond types.

A is incorrect because primary structure involves the covalent bonds between amino acids. **B** is incorrect because secondary structure involves alpha helices and beta-pleated sheets formed from the hydrogen bonds between oxygen and hydrogen atoms. **D** is incorrect because quaternary structure reflects multiple protein chains bonded together.

15 D

The proteome is the complete set of proteins produced/expressed by an organism. Glycogen is a carbohydrate, not a protein.

A, **B** and **C** are incorrect because enzymes, collagen and haemoglobin are all proteins.

16 A

On the graph, the products have more energy than the reactants, so the process must be anabolic and require/absorb energy. The activation energy is the distance between the reactants and the top of the curve.

B is incorrect because the products have more energy than the reactants and therefore will absorb energy. **C** is incorrect because the value of the activation energy from the level of R and S is to the top of the peak, not region X. **D** is incorrect because product molecules T and U have more energy than substrate molecules R and S as they are located higher on the energy axis.

17 B

Proteins are synthesised on ribosomes.

A is incorrect because translation occurs in the cytoplasm, not in the nucleus. **C** is incorrect because the Golgi body packages the protein into vesicles that will travel to the plasma membrane to be released from the cell. **D** is incorrect because ribosomes are located on rough endoplasmic reticulum (not smooth).

18 D

ALT is produced and used in the liver so, if there are high levels in the bloodstream, there are problems with the liver and the cells have broken, releasing the enzyme into the bloodstream.

A is incorrect because exocytosis is only used by cells that produce products that need to be exported for use outside of the cell. **B** and **C** are incorrect because endocytosis and phagocytosis are ways that substances *enter* the cell.

19 D

The radioactively labelled amino acids can only be tracked once the protein is made during translation. The protein will be translated on the ribosomes and the endoplasmic reticulum is responsible for modification and folding of the protein, which then goes to the Golgi apparatus for packaging into vesicles to be exported.

A is incorrect because the nucleus is not involved in translation. **B** is incorrect because the endoplasmic reticulum is involved in this process before the Golgi apparatus. **C** is incorrect because the nucleus is not involved in this process.

Short-answer questions

20 a

Nucleic acid	Sequence of nucleotides	Function of the molecule
DNA	CGA TGA CAT	DNA contains the instructions needed for an organism to develop, survive and reproduce. (1 mark)
mRNA	GCU ACU GUA (1 mark)	mRNA is a single-stranded molecule that carries a copy of the genetic code from DNA in the nucleus into the cytoplasm. (1 mark)
rRNA	-----------------------------	Ribosomal RNA associates with a set of proteins to hold the mRNA within the ribosome structure and move it along during translation. (1 mark)
tRNA	CGA UGA CAU (1 mark)	Transfer RNA carry specific amino acids on one end and a specific triplet code on the other. They carry amino acids to the ribosome for the production of a protein. (1 mark)

b i messenger RNA   **ii** transfer RNA

c Ala – Thr – Val

d Degenerate (also known as redundant) means that a single amino acid may be coded for by more than one codon. An example from the table shows that Ala (alanine) can be coded for by four codons – GCU, GCC, GCA and GCG. (1 mark) The genetic code is universal because the codon for a specific amino acid is the same in all organisms; i.e. GCU will code for alanine in bacteria, fungi, plants and animals. (1 mark)

21 a *Two of* (1 mark each)*:*

- Molecule 1 contains uracil, whereas molecule 2 contains thymine.
- Molecule 1 contains ribose sugar, whereas molecule 2 contains deoxyribose sugar.
- Molecule 1 contains more oxygen, whereas molecule 2 contains less oxygen.

Thiamine is *not* accepted because thiamine is a type of Vitamin B.

b Many nucleotide monomers contained in the DNA make up introns, which are spliced out and do not code for amino acids. (1 mark) Each amino acid is coded for by three monomers (triplet). (1 mark)

22 a In the nucleus, the DNA unwinds to expose the sequence on the template strand. RNA polymerase catalyses the production of a single strand of RNA that is complementary to the DNA template strand from free floating nucleotides. (1 mark) This is called pre-mRNA, which then undergoes posttranscriptional modification before it leaves the nucleus. Posttranscriptional modification includes the splicing out of introns and re-joining of exons, addition of a methylated cap on the 5′ end and a poly-A tail on the 3′ end. (1 mark) This processed mRNA then leaves the nucleus through nuclear pores for translation into a protein. (1 mark)

b A strand of mRNA that has left the nucleus enters the cytoplasm and attaches to ribosomes where the mRNA code is read three bases at a time (the codon). A transfer RNA molecule, carrying a specific amino acid on one end and an anticodon on the other, binds to the complementary codon on the mRNA. (1 mark) Thus, a specific sequence of amino acids is linked together through the formation of peptide bonds by condensation polymerisation to form a polypeptide chain. (1 mark) Once the mRNA sequence has been translated, the protein will undergo further processing in order for it to be folded into a functioning protein and will be packaged by the Golgi apparatus to be exocytosed from the cell. (1 mark)

> You would receive two marks for explaining the process and one mark each for the location.

23 a RNA polymerase

b *Two alternative answers were accepted here:*

- Transcription product: pre-mRNA (1 mark)
- Processing – *two of* (1 mark each): intron removal or exons joined, addition of a methyl cap/ guanine cap, addition of poly-A tail or addition of 5′ cap
- Transcription product: mRNA (1 mark)
- Processing – *two of* (1 mark each): addition of methyl cap, addition of poly-A tail or addition of 5′ cap

> Answers that said only 'capped and tailed' as the explanation received 1 mark.

c Factors expressed by regulator genes could lead to production of the different proteins. (1 mark) That is, different exons can be spliced together to form different proteins from the one DNA gene sequence. (1 mark)

> Students were required to relate their answer to how the same genetic sequence could produce different proteins.

d Universal

> A common incorrect answer was 'redundant'.

24 A repressor protein binds to the operator region on a strand of DNA, blocking the binding of RNA polymerase. (1 mark) This results in 'switching off' the gene and no transcription occurs. (1 mark)

25 Similarity: Both have a promotor region *or* both use RNA polymerase that binds to the promotor region. (1 mark)

Difference: Eukaryotes contain introns (non-coding regions) and exons (coding regions), whereas prokaryotes do not. (1 mark)

26 Eukaryotic organisms can produce multiple proteins from one gene by the splicing together of the different exons.

27 a Eight

> The final triplet on the DNA is ATT, which results in UAA on the mRNA, which is a stop codon, so no amino acid will be added for this triplet.

b AAT changed to ATT or ACT. The sixth bolded triplet needs to code for the stop codon to make a polypeptide of only five amino acids.

28 a   **i**   DNA template strand is copied during transcription or RNA polymerase joins complementary nucleotides (1 mark) and pre mRNA is formed (1 mark).

ii   Exons: are translated or are joined to form structure Q; mRNA (1 mark)

Introns: removed when splicing occurs (1 mark)

One mark was awarded if both parts were correctly named.

b   (1 mark each for:)

- Structure S is the ribosome, where the mRNA is translated.
- Structure E is tRNA, which transports a specific amino acid.
- Structure G is an anticodon, which is complementary to the codons of mRNA.
- In structure F, the product is a polypeptide.

29 a   An operon is a cluster of genes that share the same promoter and are transcribed as a single large mRNA that contains multiple structural genes.

b   Regulatory gene

c   Structural genes – *trp*E, *trp*D, *trp*C, *trp*B and *trp*A

d   Zero. Transcription would not occur. Tryptophan is present to bind to the repressor, changing its conformation. It can then bind to the operator region, stopping the transcription of the *trp* genes.

e   If there was no tryptophan present, the repressor protein could not bind to the operator. (1 mark) Therefore, RNA polymerase can bind to the promoter and transcribe the genes necessary to produce tryptophan. (1 mark)

30 a   Yes, one would result in an increase in tryptophan being produced and the other would result in none being produced.

b   Mutant X: The nucleotide change and resulting change in complementary relationship means that the repressor protein, if bound to the operator, will not dissociate. (1 mark) This results in no tryptophan being produced regardless of how much, or how little, there is in the environment. (1 mark)

Mutant Y: The change in shape of the tryptophan binding site results in tryptophan not being able to bind to the repressor protein. (1 mark) This results in there not being the conformational shape change in the repressor protein to enable it to bind to the operator region. This would lead to the uncontrolled production of tryptophan. (1 mark)

31 a   **i**   It has a carboxyl group on one side with an amine group on the other around a central carbon atom.

ii   All amino acids have a carboxyl group, an amine group and a hydrogen bonded to a central carbon. (1 mark) The fourth bond is unique to each amino acid and is called the R group or side chain. It is what defines each amino acid and can consist of a number of different arrangements of atoms. In alanine it is a CH_3, in glycine it is simply another hydrogen. (1 mark)

b   **i**   A: alpha helices (1 mark), B: beta-pleated sheets (1 mark)

ii   Hydrogen bonds

32 a   An enzyme is a biological catalyst made of protein and is not used up in the chemical reaction. (1 mark) It lowers the activation energy required for a chemical reaction and, therefore, the reaction proceeds at a faster rate. (1 mark)

b

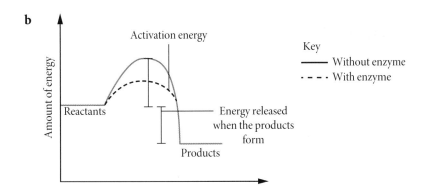

The products have a lower energy level than the reactants, which means energy is released during the reaction. (1 mark) The activation energy is the input of energy required for the reaction to proceed. This is lowered when an enzyme catalyst is used. (1 mark)

33

Protein	Function	Example
Contractile proteins	Involved in cell movement, especially muscle contraction (0.5 mark)	Actin and myosin contract in muscle movement and in the movement of cells such as phagocytes. (0.5 mark)
Hormonal proteins	Some hormones are chemical messenger proteins, which play a role in coordination of cellular processes (0.5 mark)	Insulin regulates glucose metabolism by controlling the blood sugar concentration. (0.5 mark) Growth hormone stimulates protein production in muscle cells. (0.5 mark)
Structural proteins	Fibrous proteins that provide support and strength (0.5 mark)	Keratins strengthen protective coverings such as hair, feathers and beaks. (0.5 mark) Collagens and elastin provide support in connective tissue such as ligaments. (0.5 mark)
Transport proteins	Carrier proteins that move molecules from one place to another around the body, within cells and across cellular membranes (0.5 mark)	Haemoglobin in red blood cells carries oxygen. (0.5 mark) Cytochromes are electron carriers in cellular respiration. (0.5 mark)

34 a The ribosome binds to the mRNA and tRNA brings in specific amino acids. (1 mark) The tRNA anticodon is complementary to the mRNA codon. (1 mark) The amino acids are joined by condensation polymerisation to produce the protein enzyme lipase. (1 mark)

In your answer, always remember to refer to the question being asked.

b *Three of* (1 mark each):

- Rough endoplasmic reticulum – transports lipase within the cell to the Golgi body
- Golgi body – packages the lipase into vesicles for export from the pancreatic cell
- Vesicles – membrane-bound vessels that carry lipase to the plasma membrane, where it fuses and releases the lipase from the cell
- Plasma membrane – site where the vesicle fuses and releases the contents (lipase) by exocytosis
- Mitochondria – provides the energy required for the process

RER or ER were not suitable abbreviations unless they were properly explained.

DNA manipulation techniques and applications

Multiple-choice questions

1   C

Looking at the plasmid diagram, *Eco*RI and *Bam*HI will cut once each and *Hinc*II will cut twice, resulting in four cuts and four resulting fragments.

A, **B** and **D** are incorrect because the plasmid DNA is circular, and the first cut will still only result in one fragment (as opposed to a linear piece of DNA, which, when cut, will result in two fragments).

2   B

DNA ligase binds DNA fragments together.

A is incorrect because this enzyme has nothing to do with the manufacture of antigens. **C** is incorrect because producing copies (clones) of DNA is the responsibility of DNA polymerase. **D** is incorrect because restriction enzymes cut DNA to produce blunt or sticky ends.

3   A

*Eco*RI is the only enzyme with the recognition sequence that will produce sticky ends.

B is incorrect because the recognition sequence for *Hind*III is not present in this DNA fragment (AAGCTT). **C** and **D** are incorrect because they produce blunt ends.

4   B

*Hind*III cannot cut this fragment because it does not contain the correct recognition sequence, so the fragment will only be cut once with *Eco*RI, resulting in two fragments of double-stranded DNA.

A is incorrect because the fragment is double-stranded, not single-stranded. **C** is incorrect because if the fragment was cut with *Alu*I the result would be two double-stranded DNA segments. **D** is incorrect because if the fragment was digested by all four restriction enzymes, there would be three cuts resulting in four double-stranded DNA segments.

5   B

As bacteria do not have an immune system, they contain an enzyme called CRISPR-Cas9 that cleaves viral DNA to prevent infection.

A is incorrect because this enzyme does not produce proteins of any sort. **C** is incorrect because this enzyme has no effect on plasmid replication. **D** is incorrect because CRISPR-Cas9 does not have any effect on bacterial replication.

6   D

A is incorrect because the replication of DNA during PCR is done by DNA polymerase. **B** is incorrect because PCR is often the precursor to electrophoresis. **C** is incorrect because gene cloning is a process where genes are copied and amplified and does not involve CRISPR-Cas9.

7   C

Using the DNA ladder on the left as a guide, sample Y is the only one with 4 bands in the correct locations.

A, **B** and **D** are incorrect because, although they all have four bands, they are in the incorrect locations when compared to the known fragments in the DNA ladder in Lane S.

8   B

The negative electrode is the one closest to the wells. This is because DNA is negatively charged and is attracted to the positive electrode, hence the molecules will migrate through the gel towards it. Sample X has a band that is closest to the wells.

A, **C** and **D** are incorrect because they have bands that have moved further down the gel towards the positive electrode than sample X.

9 A

PCR uses specific primers to amplify sections of DNA.

B is incorrect because the primers are specific for one pathogen. **C** is incorrect because bacteria (along with their plasmids) are used in recombinant DNA technologies, not PCR. **D** is incorrect because restriction enzymes are used to cut DNA in recombinant DNA technologies.

10 C

The restriction enzyme *Sal*I will cut this strand three times, creating four fragments; therefore, four bands on an electrophoresis gel.

A and **B** are incorrect because there are three fragment bands in each. **D** is incorrect because there are six bands.

11 A

PCR detects and amplifies specific DNA sequences specific to the pathogen, so results will be either positive or negative depending on whether the virus is present.

B is incorrect because antibodies are proteins and not involved in the PCR process. **C** is incorrect because gel electrophoresis can be used after PCR if real time PCR is not used. **D** is incorrect because plasmids are involved in recombinant DNA technologies.

12 D

Anthocyanin is only produced by bacteria 4.

A, **B** and **C** are incorrect because the products from the previous bacterial culture are the reactants for the next.

13 B

This is a complex process that requires a staged approach to the production of anthocyanin. Only bacteria 4 produce the anthocyanin, so large quantities of this bacteria are required.

A is incorrect because restriction enzymes are used to produce the recombinant plasmids before they are used to transform the bacteria. **C** is incorrect because the number of genetic instructions required is too large for just one bacterium to produce anthocyanin efficiently. **D** is incorrect because bacteria do not discriminate in terms of which plasmids they accept during transformation.

14 C

Antibiotic resistant genes included on the transformed plasmid can make identifying transformed bacteria easy due to their abilities to grow in the presence of the antibiotic, whereas non-transformed bacteria will be sensitive to the antibiotic and will not grow.

A is incorrect because only bacteria 4 will produce anthocyanin when supplied with the correct reactant. **B** is incorrect because not all bacteria are transformed during the transformation process, so a means of selection must be implemented. **D** is incorrect because this process is like a biochemical pathway where the products of one reaction are the reactants for the next, so mixing all four bacteria together will not provide the selection required.

15 B

GMO is an overarching term used to describe an organism that has had its genome manipulated in some way. This can include a number of different techniques such as silencing genes, multiplying genes, inserting genes or controlling the expression of gene products.

A is incorrect because it describes a mutation and may not result in any change to the protein. **C** is incorrect because it is describing artificial selection. **D** is incorrect because GMOs can modify the genomes of other organisms.

16 C

A transgenic organism is a type of GMO where the genes from another species are introduced into an organism.

A is incorrect because adapting to a new environment is natural selection. **B** is incorrect because transgenesis requires genes from a different species. **D** is incorrect because exposure to a mutagen will cause a random change to the genome of an organism, so will not produce the desired result.

17 C

> Negative impacts on ecosystems could be reduced because fewer insecticides were needed to control insects over time.
>
> **A** is incorrect because, looking at the graph, the line for the insecticide use goes down over time and the line for the Bt corn grown (percentage area) goes up. **B** is incorrect because the question includes no mention of cost. Also, an assumption could be made that any genetically modified crop will have had significant money invested in its creation that needs to be re-cooperated by the organisation and/or the crop could be trademarked. **D** is incorrect because if the trend line from the graph is to continue, more farmers will plant Bt corn in the future.

Short-answer questions

18 a *Hind*III produces sticky ends whereas *Hae*III produces blunt ends.

   b DNA polymerase is used in the replication of DNA such as the extension of nucleotides in a DNA strand during PCR. DNA ligase mends the phosphodiester-deoxyribose backbone to join pieces of DNA together. (1 mark) The result of the error is that the researcher will not produce recombinant DNA because DNA polymerase will not catalyse this reaction. (1 mark)

   c No, the plasmid and genome must be cut with the same restriction enzyme (1 mark) so that the ends are complementary, especially in the case of restriction enzymes that create sticky ends such as *Eco*RI and *Bam*HI. (1 mark)

19 a   **i** The guide RNA is a sequence of nucleotides used to guide the enzyme Cas9 to the correct position on the target DNA sequence.

      ii It is important that Cas9 endonucleases only cut the viral DNA and not the bacteria's genome. In order to protect its own DNA, it will only cleave at sites where a PAM (protospacer adjacent motif) is present. (1 mark) A protospacer is a 2–6 base sequence that immediately follows the target sequence for Cas9. In the bacteria *Streptococcus pyogenes*, Cas9 recognises the PAM NGG, where N can be any nucleotide followed by two guanine bases. Cas9 will cut where this sequence is found directly following the gene of interest, therefore protecting its own genome that does not have the PAM sequence, while at the same time cutting viral DNA. (1 mark)

   b Scientists can edit genes by synthesising a single guide RNA (sgRNA), which targets a specific section on the DNA. (1 mark) Cas9 will target this sequence if the correct PAM is present and cleave the DNA. (1 mark) Once Cas9 has bound to the specific section of DNA and has cut it, the gene sequence can then be edited by inserting the desired piece of DNA. (1 mark)

20 Peanut allergies are one of the most common allergies in humans and can result in anaphylaxis, a life-threatening condition. (1 mark) Producing peanuts that do not contain the proteins that trigger this anaphylactic response in humans could save people's lives. (1 mark)

21 a Denaturation (95°C): The DNA is heated to break the hydrogen bonds between the complementary bases along the DNA strand. (1 mark)

    Annealing (55°C): The DNA is cooled, and primers attach to the DNA and promote the replication process from the point of attachment. (1 mark)

    Extension (72°C): The DNA polymerase known as Taq polymerase extends each strand beyond the primer, adding the free nucleotides to replicate each DNA strand. (1 mark)

    This is repeated multiple times so that, each time the cycle is completed, it will double the quantity of DNA. (1 mark)

   b The primers selected are specific for a DNA sequence in the pathogen. (1 mark) This means that if the pathogen is present, the DNA will be amplified, and a positive result will be obtained. If the pathogen is not present, no DNA amplification will occur. (1 mark)

22 a Short tandem repeats

   b Individuals have a unique number of repeats with the same sequence, so they will produce unique fragment profiles based on the size of the sequence.

23 a 750 bp and 1700 bp

b Suspect 3, (1 mark) because the DNA from the crime scene matches the sample taken from Suspect 3. The bands are the same sizes within the gel electrophoresis. (1 mark)

c *Two of* (1 mark each): temperature, voltage applied, gel concentration, ionic strength of the buffer

24 a No. This is a random process, so it cannot be guaranteed that the desired gene has been inserted into the plasmid. Other events may have occurred, such as the plasmid joining with another plasmid, or the plasmid simply re-joining the ends cut by the restriction enzyme.

b No. The plasmids need to be in close proximity to the bacteria in solution and the bacterial cell membrane has to have been affected by the heat shock treatment. This is also a random process.

c The plasmid should also include an antibiotic resistant gene. The bacteria that have been transformed will now be resistant to the antibiotic and, therefore, will grow in a nutrient solution containing that antibiotic.

d The transformed bacteria replicate the plasmids, which then produce the required gene product. In addition, bacteria multiply rapidly and therefore produce the protein in large quantities. (1 mark) Bacteria are a good choice because they can be cultured in large volumes and the desired protein can be produced in a short period of time. (1 mark)

25 a Diabetes is one of the most common diseases in Australia and many people require insulin injections several times a day. Insulin can be obtained from the pancreas of pigs and cows, but insulin made from recombinant DNA is a more efficient process to produce the large quantities required.

b A plasmid is a circular piece of DNA (1 mark) and it is called a vector because it can be a transporter of a target piece of DNA into a host cell (1 mark).

c D, B, G, E, A, F, C

d DNA ligase

26 a Both are transgenic organisms because transgenic organisms have genes from a different species. (1 mark) Bt cotton has genes from *Bacillus thuringiensis* (1 mark), while golden rice has genes from *Erwinia uredovora* and *Narcissus pseudonarcissus* (1 mark).

b Reducing insect damage means that the cotton plants will grow more and produce more cotton, therefore increasing crop yield.

c *Several responses have been given. Only one implication is required.* (1 mark per implication)

	Social implication	Biological implication
Bt cotton	• Farmers cannot use saved seeds; therefore, it is expensive because they need to buy new seeds each year • May cause skin disease, reducing quality of life for farmers • Less money available as need to buy seed, which impacts on other areas of life, such as food, education etc. • Could increase yield, resulting in more money and potential increases in standard of living	• Might contribute to skin diseases in farmers • Reduces the environmental impact of pesticides • Genes might get into weed crops and reduce the number of predators for insects that feed on Bt cotton
Golden rice	• Farmers can save rice to be used in the next year's harvest, which can lead to increased profits • Golden rice improves socioeconomic levels because of a reduction in death and disease • Reduced deaths may lead to decreased birth rates • Proven safe	• Improved nutrition through increased intake of Vitamin A • Saves lives as avoids Vitamin A deficiency • Increased intake of Vitamin A improves health, leading to benefits to the community

27 a CRISPR-Cas9, transformed bacteria, or a viral vector could be used to insert genes into plant cells. (1 mark) *or* Biolistics could be used, where a gene gun is used to directly shoot DNA fragments into a plant cell. (1 mark)

    b *Experimental design should include:*

- Identification of the dependent and independent variable (IV – Salt concentration of the soil OR water, DV – height of plant, greater production of tomatoes or rate of growth over time) (1 mark)

- Method, including two groups of tomato plants: experimental group with 200 seeds from transformed plants and control group with 200 seeds from wild type plants. All other variables are controlled. All seeds are grown under varying percentages of salt concentration (IV) and the DV measured after the same time period. (2 marks)

- Results that would be expected in plants that had been successfully transformed; e.g. greater tolerance of higher salt concentrations as evidenced by increased height. (1 mark) Valid experimental design with multiple trials. (1 mark)

CHAPTER 2, UNIT 3 AREA OF STUDY 2

Regulation of biochemical pathways in photosynthesis and cellular respiration

Multiple-choice questions

1 D

A biochemical pathway is a cascade of reactions where the products of one reaction are the reactants of the next. **A**, **B** and **C** are incorrect because they are all true statements.

2 B

ATP is a coenzyme, fructose 6-phosphate is the substrate and phosphofructokinase is an enzyme and therefore will increase the rate of the reaction.

A is incorrect because ATP is a coenzyme. **C** is incorrect because fructose 6-phosphate is the substrate. **D** is incorrect because ATP is used in the reaction, not released as ADP.

3 A

Coenzymes are non-protein organic molecules that cannot catalyse reactions by themselves but are necessary for enzyme-catalysed reactions to proceed.

B is incorrect because coenzymes are non-protein organic molecules. **C** is incorrect because coenzymes bind to the enzyme not the reactants. **D** is incorrect because coenzymes cannot catalyse reactions alone.

4 C

C is the only option with valid molecules for this process. **A**, **B** and **D** are incorrect because CTP, NAP and NAH are not involved in these cellular processes.

5 C

Optimum activity for an enzyme is shown by the peaks on the graphs. Enzyme W will be most active at pH 3 and 37°C. Enzyme Y will be denatured at temperatures above approximately 45°C and enzyme Z cannot be an intracellular human enzyme because its optimum range is 75°C, much higher than the human body temperature of 37°C.

A is incorrect because at pH 7 enzyme Y has reduced activity when temperature is decreased it does not denature. **B** is incorrect because enzyme Z has an optimum temperature of 75°C, whereas the human body functions optimally at 37°C. **D** is incorrect because enzyme X has an optimum pH 7 at 37°C.

6 D

The shapes of the graphs are very important for the variables investigated. Temperature and pH will have a bell curve, whereas substrate and enzyme concentration will increase and plateau at some point. Variable 1 indicates that A and C are incorrect. Variable 2 shows **D** is the correct answer.

A, **B** and **C** are incorrect because Variable 1 and Variable 2 will both be temperature or pH.

7 A

The active site of an enzyme has a specific conformation (shape) that can be influenced by changes in the secondary and tertiary structures. Heat interrupts the bonds between amino acids from different parts of the chain and influences the folding of the protein.

B is incorrect because heat changes the bonding and folding of the protein, not the structure of the amino acid monomers. **C** and **D** are incorrect because the amino acid sequence is not changed.

8 B

A competitive inhibitor will bind to the active site of the enzyme, so the shape of the inhibitor must match the active site exactly.

A, **C** and **D** are incorrect because they do not have the complementary shape to the active site.

Short-answer questions

9 a In a biochemical pathway, the products of one reaction are the reactants of the next. If no reactants are supplied, the whole biochemical pathway will cease, thus reducing the levels of NADH produced throughout the whole pathway and affecting energy production.

  b **i** The biochemical pathway will cease at that point in the pathway.

    ii No. For the bloom to look healthy for a long period of time, the cells must be able to produce ATP for cellular processes to continue. (1 mark) If the pathway is interrupted by a malformed enzyme, the cells will no longer be able to provide ATP for other processes and the bloom will die. (1 mark)

10 ATP-synthase is an enzyme that is embedded in the membranes of the mitochondria (1 mark), which is integral in the production of ATP by adding inorganic phosphate to a molecule of ADP (1 mark).

11 a pH 2 is highly acidic. This will denature the enzyme, resulting in a conformational change in the active site. This will make the enzyme unable to catalyse the first reaction in the pathway.

  b A decrease in temperature takes the enzyme out of its optimum range and so decreases the enzyme's activity. This will result in a slowing down of the reaction rate.

A decrease in temperature does not denature the enzyme. The lower temperature reduces the kinetic energy of the molecules and as a result there are fewer enzyme–substrate collisions occurring.

12 a **i** The production of ATP and alanine would continue unregulated, resulting in an uncontrolled production of products.

    ii If the inhibition was not reversible, the enzyme would be permanently inhibited, resulting in the enzyme no longer functioning and products no longer being synthesised.

  b **i** D

    ii Diagram B shows competitive inhibition where the inhibitor binds to the active site of the enzyme, resulting in competition for the active site between the inhibitor and the substrate. (1 mark) Diagram D shows non-competitive inhibition, where the inhibitor binds to another site on the enzyme, changing the conformation of the active site and resulting in substrates no longer being able to bind for catalysis. (1 mark)

13 a **i** Tube 1 – higher concentration of enzyme, therefore faster rate

    ii Tube 3 – lower concentration of starch at the beginning of the reaction

    iii Tube 4 – Tube 5 was the control and did not include any enzyme

b Measurements would need to be taken at timed intervals.

c i

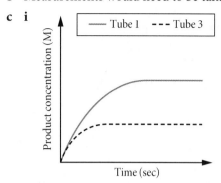

ii

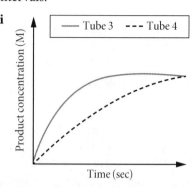

Tubes 1 and 3 have the same rate because they have the same amount of enzyme (so will rise at the same gradient, assuming saturation of the enzyme) (1 mark) but Tube 3 will result in half of the products produced due to having half the concentration of starch initially. (1 mark)

Tube 3 will have a steeper gradient because it has a higher concentration enzyme than Tube 4 (1 mark). They will, however, still end up at the same level because they both have the same concentration of starch, Tube 4 will just take longer (have a slower rate). (1 mark) The gradient of the graph indicates rate of the reaction.

Photosynthesis as an example of a biochemical pathway

Multiple-choice questions

1 B

The light-independent reaction occurs in the stroma as indicated by U.

A is incorrect because T indicates the outer membrane of the chloroplast. **C** is incorrect because V indicates a piece of nucleic acid. **D** is incorrect because W indicates ribosomes.

2 D

Outputs of the light-independent stage include water, $NADP^+$, ADP, inorganic phosphate and glucose.

A is incorrect because carbon dioxide is an input of photosynthesis. **B** is incorrect because ATP is a product of cellular respiration. **C** is incorrect because oxygen is a product of the light-dependent stage of photosynthesis.

3 B

The important parts of the diagrams to look at are the coenzymes produced/used in the reactions. The arrows in the middle of each diagram indicate the direction of the reaction, relative to the left or right reaction.

A is incorrect because NADPH is used in photosynthesis. **C** is incorrect because it has one of the reactions reversed. **D** is incorrect because glucose (carbohydrate) and water are an output in the light-independent stage.

4 D

NADPH is involved in photosynthesis.

A, B and **C** are incorrect because glycolysis, the Krebs cycle and electron transport chain in mitochondria are all stages of cellular respiration.

5 A

The splitting of water occurs in the grana of the chloroplast during the light-dependent stage. This forms oxygen ions that then form oxygen gas and hydrogen ions, which can then be accepted by $NADP^+$ to become NADPH.

B is incorrect because the Calvin–Benson cycle is the light-independent stage and occurs in the stroma. Carbon dioxide is also a reactant, not a product. **C** is incorrect because the electron transport chain occurs in the thylakoid membranes of the grana but does not use carbon dioxide. **D** is incorrect because glucose is formed in the stroma of the chloroplast.

6 B

C_4 ('four-carbon') and CAM plants initially attach CO_2 to PEP (phosphoenolpyruvate) to form the four-carbon compound OAA (oxaloacetate) using the enzyme PEP carboxylase. This efficiently stores the carbon for later processing and minimises water loss. Plants do not complete all stages of photosynthesis at night. They can rearrange how they process the reactants to minimise water loss and close their stomata during the day.

A is incorrect because plants close their stomata during the day. **C** is incorrect because the light-independent stage of photosynthesis cannot occur at night. **D** is incorrect because the difference is in the changing of the processing of carbon dioxide, not limiting oxygen.

7 A

C_3 plants are the most abundant type of plants on the globe and only use the one photosynthetic pathway to produce glucose and oxygen.

B is incorrect because they are the most abundant plants on the globe. **C** is incorrect because they only use the one photosynthetic pathway. **D** is incorrect because they only produce two main photosynthetic products: glucose and oxygen.

8 A

To increase the rate of photosynthesis, you need to either increase the concentration of the reactants (carbon dioxide and water), decrease the concentration of the products (glucose and oxygen) or increase the temperature.

B is incorrect because decreasing the concentration of carbon dioxide will decrease the rate. **C** and **D** are incorrect because increasing the concentration of oxygen and glucose (products of photosynthesis) will decrease the rate.

9 C

Carbon dioxide is used in the Calvin–Benson cycle (light-independent stage) and, when there is more substrate, the rate will increase.

A is incorrect because the Calvin–Benson cycle is the light-independent reactions and rate does not increase with a longer day. Photosynthesis occurs for a longer time, but the rate remains constant; light intensity has to increase to increase the rate. **B** is incorrect because moving a plant from the sunlight into shade reduces the light intensity and therefore reduces the rate of photosynthesis. **D** is incorrect because the stomata in a plant usually close during the night when photosynthesis often slows due to the absence of light.

10 C

For fast growth rate, plants need ample warmth, sunlight, water and the reactants of photosynthesis.

A is incorrect because Iceland has very cold temperatures and rates of reactions will be slower. **B** is incorrect because deserts are extremely hot, so plants will take measures to decrease water loss, which will affect growth rate. **D** is incorrect because 3 cm in 10 years is extremely slow.

Short-answer questions

11 a Grana or thylakoid (membrane)

b

Name of the stage of photosynthesis that occurs at X	Light-dependant stage (1 mark)	
Two input molecules that are required for reactions at X	1 Water (*light was not acceptable*) (0.5 mark)	2 *One of* (0.5 mark): • ADP • P_i • NADP
Two output molecules that result from the reactions at X	1 Oxygen (0.5 mark)	2 *One of* (0.5 mark): • ATP • NADPH or H^+ (*NAD was not acceptable*)

The question asked for molecules, so light was incorrect. Some students also incorrectly used NAD and NADH, which were not acceptable.

12 a **i** Water and carbon dioxide. Chemical symbols were equally acceptable.

ii Water is split in the light-dependent reaction to produce oxygen gas. (1 mark) Oxygen produced diffuses into the stroma (region R). *or* When light is not available, oxygen is not produced. (1 mark)

b NADPH transfers hydrogen ions. *Protons and electrons were also an acceptable answer.* (1 mark) ATP transfers energy. (1 mark)

Students needed to state the role of each enzyme to achieve full marks.

13 a Sample 3, Wavelength 3

b Green light is reflected by plants and most of the chlorophyll in plants respond to red and blue light; therefore, oxygen will be produced in very small amounts.

14 a Hot, dry areas such as deserts

b Grassland areas

c Close their stomata during the day, take up CO_2 at night when the air temperature is lower, or concentrate CO_2 in their tissues

15 The reactants in photosynthesis are carbon dioxide and water. When one of these reactants is limited, the reaction slows down. (1 mark) The lack of water slows the rate of photosynthesis, which in turn slows the rate of plant growth, reduces leaf size and amount, and results in less fruit and changes in reproductive phases. (1 mark) In introduced species, this results in crop losses and death of the plants because they are not adapted to this environment. (1 mark) Australian native plants have adapted to these conditions to ensure the survival of the species. Adaptations such as dormancy or changing the time of day when stomata open can reduce water loss and enable the Australian native plants to survive. (1 mark)

16 a $6CO_2 + 12H_2O \rightarrow C_6H_{12}O_6 + 6O_2 + 6H_2O$

b **i** Some scattered light was able to get in underneath the foil cover (1 mark) and photosynthesis continued for a short time after the leaf was covered, which would have incorporated the radioactively labelled carbon (1 mark).

ii The temperature of the leaves was at 70°C, which would have denatured the enzymes for the biochemical pathway of photosynthesis. (1 mark) This would result in the pathway stopping and no photosynthesis occurring once the plant cells reached 70°C. (1 mark)

Cellular respiration as an example of a biochemical pathway

Multiple-choice questions

1 A

All living cells require mitochondria to produce energy for cellular processes.

B, **C** and **D** are incorrect because all cells use cellular processes to survive, and thus require energy and mitochondria to undertake them.

2 C

The cristae of the mitochondria is where the enzyme ATP synthase is located, which is where the majority of ATP is produced in the mitochondria.

A is incorrect because glycolysis occurs in the cytosol of the cell. **B** is incorrect because NAD^+ is converted into NADH during the Krebs cycle, which occurs in the mitochondrial matrix. **D** is incorrect because pyruvate enters the mitochondria from the cytosol just before the start of the Krebs cycle.

3 D

Respiration uses glucose and oxygen to form carbon dioxide and water. During respiration, oxygen is used in the electron transport chain and combines with the hydrogen to form water. The radioactive label will therefore be incorporated into the water molecules.

A is incorrect because oxygen is not incorporated into adenosine triphosphate (ATP). **B** is incorrect because the oxygen in the carbon dioxide comes from the breakdown of glucose. **C** is incorrect because glucose is a substrate, not a product of cellular respiration.

4 A

The loaded coenzyme NADH, ADP and inorganic phosphate enter the electron transport chain where electrons are transferred through a series of reactions until they are finally accepted by oxygen. This sets up a proton gradient that is used to drive the formation of ATP. This process also produces water and the unloaded coenzyme NAD^+.

B is incorrect because oxygen is an input and water is an output. **C** is incorrect because the loaded coenzyme NADH is an input. The hydrogen ions from this coenzyme are pumped into the mitochondrial intermembrane space and then flow down a concentration gradient into the mitochondrial matrix to drive the production of ATP. **D** is incorrect because $NADPH/NADP^+$ are used in photosynthesis. Also, oxygen is an input and water is an output.

5 C

Anaerobic fermentation in yeast cells produces ethanol and carbon dioxide. The reactant is glucose, lactic acid is produced via anaerobic respiration in animal cells and water is produced in aerobic respiration.

A is incorrect because glucose is a substrate. **B** is incorrect because lactic acid is the product of anaerobic respiration in animal cells. **D** is incorrect because water is the product in aerobic cellular respiration.

6 A

Anaerobic respiration produces ATP.

B, **C** and **D** are incorrect because they all require energy and hence use ATP. This results in a net loss of ATP molecules.

7 D

A decrease in the concentration of glucose will decrease the rate of cellular respiration because it is a substrate.

A is incorrect because increasing the concentration of a reactant increases rate. Oxygen is a reactant in cellular respiration. **B** is incorrect because an increase in temperature increases the rate of reaction. **C** is incorrect because decreasing the concentration of a product in a reaction increases rate. Carbon dioxide is a product of cellular respiration.

8 B

At point T there is zero net output of oxygen, so what is produced is being used by cellular respiration.
A is incorrect because there is a negative output of oxygen, which means the rate of cellular respiration is higher than the rate of photosynthesis. **C** is incorrect because more oxygen is being produced than is being used. **D** is incorrect because the graph has plateaued at just under 0.2, meaning oxygen is being produced but another limiting factor is resulting in a constant rate of production despite higher light intensities.

9 B

The rate of oxygen output has become constant despite an increase in light intensity, so there is a limiting factor involved. This could be concentration of reactants or saturation of the enzymes involved in the biochemical pathway.

A is incorrect because the rate has become constant, not stopped. **C** is incorrect because at point T the rate of photosynthesis is equal to the rate of respiration. **D** is incorrect because the optimum light intensity is 50 AU. Light intensities above this are not going to increase rate any further.

Short-answer questions

10 a Glycolysis

b Ethanol should be included because the products of anaerobic cellular respiration depend on the cell type.

c Mitochondria

d

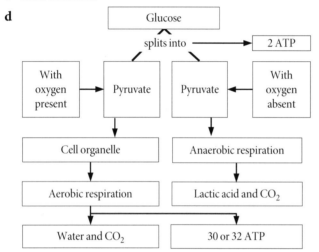

11

Biochemical process	Location	Inputs	Outputs	No. of ATP produced
Glycolysis	Cytosol of the cell	Glucose, NAD⁺	Pyruvate, NADH	2 ATP
Krebs cycle	Mitochondrial matrix	Pyruvate	Carbon dioxide, NADH	2 ATP
Electron transport chain	Mitochondrial christae	NADH, oxygen, ADP, inorganic phosphate	NAD⁺, water	26 or 28 ATP

You would receive two marks per process.

12 a Ethanol levels rose because ethanol is a product of anaerobic respiration. (1 mark) Oxygen levels decreased because oxygen is required for aerobic respiration. (1 mark)

'Cellular respiration' was not a suitable term to use in the answer.

b Prediction: Increase (1 mark)

Explanation: CO_2 is a product of *(one of)* cellular respiration, aerobic respiration, anaerobic respiration. (1 mark)

'Respiration' alone was not awarded any marks.

c i Light-independent stage or Calvin–Benson cycle.

Many students initially gave the light-independent stage but changed their answer to the light-dependent stage.

ii

Name of input	Role
ATP	Provides energy to form glucose (1 mark)
NADPH	Carries hydrogen ions (1 mark)

NAD was incorrect because it is a carrier molecule in cellular respiration and not in photosynthesis. 1 mark for both input and its role.

13 a In the presence of both oxygen and glucose for 3 hours

b i During cellular respiration in the production of ATP

ii During respiration, inorganic phosphate is added to ADP to form ATP in both glycolysis and the Krebs cycle.

c i Anaerobic cellular respiration

ii $C_6H_{12}O_6 \rightarrow CH_3CH_2OH + CO_2 + 2$ ATP

d i Aerobic cellular respiration

ii $C_6H_{12}O_6 + 6O_2 \rightarrow 6H_2O + 6CO_2 + 30$ or 32 ATP

e Aerobic respiration creates significantly more ATP than anaerobic respiration. (1 mark) Therefore, tube 3 will use significantly more phosphate, creating 30 or 32 ATP as opposed to 2 ATP. (1 mark)

14 a In the cytoplasm of the cell

b 2 ATP

c During anaerobic fermentation, only 2 ATP molecules are created, whereas aerobic respiration creates 30 or 32 ATP molecules. (1 mark) Therefore, aerobic respiration is the more efficient process. (1 mark)

15 a Dion's muscles needed more energy due to the increased demand, so the rate of cellular respiration had to increase to meet that demand. (1 mark) This resulted in him puffing hard and increasing the concentration of oxygen in his blood stream, and decreasing the concentration of carbon dioxide in order to increase the rate of aerobic cellular respiration. (1 mark)

b Cellular respiration is a catabolic reaction, so it releases energy. This makes Dion's temperature rise. (1 mark) The optimum temperature for cellular respiration is 35°C and the extremities can dip below these temperatures, so this heat can place the enzymes into their optimum range. (1 mark)

c The enzymes would be outside their optimum range and they could potentially denature, resulting in decreased efficiency of cellular respiration and other biochemical pathways.

d Dion could not increase the concentration of oxygen high enough in his blood stream to meet the energy demands of his muscles. (1 mark) So his body switched to anaerobic cellular respiration to try to reach the demands. This resulted in the production of lactic acid, which built up in his muscles. (1 mark)

e glucose → lactic acid + 2 ATP

Biotechnological applications of biochemical pathways

Multiple-choice questions

1 A

As human populations increase around the world, the demands for food are higher. Also, changing climates can threaten the production of crops.

B is incorrect because there is really no need to grow blue tomatoes. **C** is incorrect because increased growth rate is preferable. **D** is incorrect because it is desirable for crops to take less time to mature.

2 D

Improving the soil is a different process and does not require the modification of the plant using CRISPR-Cas9.

A, **B** and **C** are incorrect because they are all valid reasons to improve crop productivity.

3 D

Modifying animals does nothing to combat the rising levels of carbon dioxide in the atmosphere or reduce the land required for crops.

A, **B** and **C** are incorrect because they are all solutions to the problems posed in the question.

4 A

Biofuels include biogas, bioethanol and biodiesel, where fibrous plant and animal waste can be used to create hydrocarbons that can be burnt as a fuel.

B is incorrect because it is the waste from animals that can be used, not the animals themselves. **C** is incorrect because farmers could use diesel in their tractors derived from fossil fuels. **D** is incorrect because biofuels are generated from waste and combusted for energy, not produced within living organisms.

5 C

Anaerobic processes occur in the absence of oxygen, so increasing the oxygen supply for bacteria is not an anaerobic process.

A, **B** and **D** are all incorrect because they are all anaerobic processes.

6 B

There are many different strains of algae that can photosynthesise. This results in a high biomass quite quickly because they are fast-growing.

A is incorrect because microalgae grow quite quickly. **C** is incorrect because they do photosynthesise. **D** is incorrect because there are many different strains.

Short-answer questions

7 **a** The use of CRISPR-Cas9 is cheaper, easier, and more selective and efficient than the use of transgenesis. It allows scientists to edit multiple targets simultaneously, which results in what is known as 'trait stacking' and manipulation of complex gene networks related to things such as drought tolerance. (1 mark) Transgenesis can be expensive and time consuming and there is a limited amount of genetic material that can be transferred at one time. (1 mark)

   b Accelerated evolution may provide an efficient pathway to achieving higher agricultural productivity and food security (1 mark) compared with the slower adaptations that occur in the natural environment. It is essential to achieve these changes in a shorter period of time in the face of population growth, global warming and climate change. (1 mark)

8 **a** Biogas (1 mark) and bioethanol (1 mark)

   b $C_6H_{12}O_6 \rightarrow 2CO_2 + 2C_2H_5OH$ (1 mark)

   Glucose → carbon dioxide + ethanol (1 mark)

   c Sustainable – Raw materials are readily available such as algal, plant or animal waste, and can be replenished in a short period of time. (1 mark)

   Carbon neutral – Carbon dioxide is taken out of the environment and used in the structural components of the plant (cellulose), structural components of animals (tissues and organs) OR undigested fibre excreted by animals. This waste can then be fermented and burnt as fuel, resulting in the carbon dioxide returning to the atmosphere resulting in no net increase to atmospheric levels. (1 mark)

d   (4 marks)

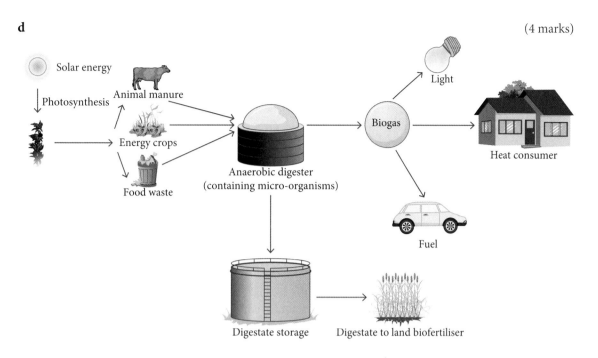

9 a *Two of* (1 mark each): burning of methane for cooking and heating, mixing bioethanol with other fuels to reduce overall fuel costs, use of methane or ethanol in fuel cells to generate electricity or power generators for electricity production

  b *Three of* (1 mark each): Reduced fuel costs, therefore increase profit; biomass that was considered waste can now be used to generate an income; reduction in the reliance on fossil fuels; sustainable practices as the reactants can be replenished in a short time frame; no net increase in carbon dioxide emissions (carbon neutral)

  c Competition for the growth of crops for food or fuel. (1 mark) With a growing population, more food production will be needed, not less. (1 mark) Climate change is a factor. (1 mark) If more land is needed for crop production, more forests will have to be removed, resulting in a larger impact on climate change and impacts on natural habitats. (1 mark)

10 a   i Raw reactants: Livestock waste, crops, food and water waste (2 marks)
     ii Uses: Fertiliser, soil amendments, fuel, gas grid (2 marks)

   b Anaerobic conditions, pH 7, 37°C, there must be an inlet for raw materials, there must be a mechanism to trap the gases or ethanol produced (3 marks)

CHAPTER 3, UNIT 4 AREA OF STUDY 1

Responding to antigens

Multiple-choice questions

1 A

Intact skin is the only option that prevents entry.

B is incorrect because normal flora provide competition for resources and space to help inhibit growth. **C** is incorrect because macrophages engulf pathogens once they have gained entry. **D** is incorrect because stomach acid destroys some pathogens but not all.

2 A

Marigolds are often used as companion plants to fruits and vegetables because they produce toxic chemicals that keep parasites away from other plants.

B is incorrect because plants do not have mobile immune cells like macrophages. **C** is incorrect because plants do not produce antibodies. **D** is incorrect because a waxy leaf surface can act as a physical barrier to prevent the pathogen from coming into contact with the plant cells.

3 A

There is a reduced number of infections in the warm, wet season when the frog's skin is saturated with moisture and a large number of beneficial bacteria are present. In the dry season, these bacteria are in lower numbers.

B is incorrect because fungi thrive in warm, wet conditions and will not rot under these conditions. **C** is incorrect because fungi mostly prefer warm, wet conditions. **D** is incorrect because the frog would be in its aquatic environment so the washing away of spores is not a significant factor in this scenario.

4 D

Interferon can be released by virally infected cells as a signalling molecule to other cells in the area.

A is incorrect because the inflammatory response is not part of the T cell defence. **B** is incorrect because histamine is released by mast cells. **C** is incorrect because platelets are involved in the clotting process and do not adhere to virions.

5 C

Dendritic cells are phagocytic antigen-presenting cells that act as messengers between the innate and specific immune responses. As such, they can present non-self antigens to lymphocytes.

A is incorrect because they are not part of the specific immune response only; they are the link between innate and specific immune systems. **B** is incorrect because they are produced in the bone marrow. **D** is incorrect because they are not involved in the inflammatory response.

6 C

Capillaries become more permeable to allow the required immune system components to gain easier access to the site of potential infection.

A is incorrect because mast cells release histamine. **B** is incorrect because the skin would become hot and red due to increased permeability of the capillaries. **D** is incorrect because phagocytes such as macrophages would engulf foreign material.

7 A

In parasitic infections, eosinophils increase in the blood stream to help combat this type of infection by releasing substances that kill parasite larvae. They are also a part of the innate immune response.

B and **C** are incorrect because helper T cells and B memory cells are part of the specific immune response. **D** is incorrect because parasites like hookworm are too large for dendritic cells to phagocytose.

8 B

Membrane receptors are on cells and come into contact with antigens that are either self antigens or non-self antigens.

A is incorrect because antibodies are secreted by B lymphocytes in response to a specific pathogen. **C** is incorrect because antigens are recognised by cell receptors. **D** is incorrect because lipoprotein function is to transport hydrophobic lipid molecules in water.

9 D

Epithelial cells are the body's cells that line external surfaces and display self antigens.

A and **B** are incorrect because bacteria and viruses are not self cells. **C** is incorrect because sperm cells are male cells only.

10 D

Look at the features of each and their size. X is a virus so will be reproduced by the invaded host cells.

A is incorrect because pathogen Z is a prion made of protein. **B** is incorrect because W is cellular, as it is bacteria, but X is not cellular, as it is a viral particle. **C** is incorrect because Y is a parasite so does not reproduce by binary fission as it is too complex and multicellular.

Short-answer questions

11 *Two of* (1 mark for each barrier; 1 mark for each explanation)*:*

- Mucous membranes (chemical barrier) – sticky mucous on the lining of the respiratory tract traps pathogens so they do not come in contact with respiratory tract cells and can then be expelled.
- Cilia (physical barrier) – hair-like projections on respiratory cells 'sweep' mucous and debris up the trachea to be expelled or swallowed.
- Commensal bacteria in the throat (microbiota) – these beneficial bacteria colonise the throat and prevent infection by producing antimicrobial substances and competing for nutrients and adhesion sites.

> The question is referring to the respiratory tract. Make sure your response is specific to the part of the body mentioned in the question.

12 a *Examples of chemical barriers in plants include* the secretion of a toxin or an odour that is harmful or unfavourable to pathogens, or the production of enzymes that affect pathogen functioning. (1 mark for one barrier)

   Examples of physical barriers in plants include waxy leaf coatings, thick bark of the stem, hairs and thorns or sunken stomata. (1 mark for one barrier)

> Students often struggle to complete questions involving plants. Make sure you do not ignore them in your studies, particularly regarding the entry of pathogens.

   b *Examples of physical barriers in humans include* intact skin, respiratory, gastrointestinal and genital tracts lined with mucous or hairs to trap pathogens. (1 mark for one barrier)

   Examples of chemical barriers include lysozymes in tears and saliva, sweat and oil secretions on the skin, low pH of the vagina or highly acidic conditions in the stomach. (1 mark for one barrier)

> Make sure you include 'intact' in your response for 'intact skin'. Skin alone does not prevent entry of pathogens. You need to be specific.

13

Cell type	Function
Macrophage	Phagocytic cell that engulfs pathogens and is found throughout the body tissues; an antigen-presenting cell that can link the innate and specific immune response (1 mark)
Neutrophil	White blood cell found in blood and tissue that rapidly enters sites of inflammation, engulfing pathogens; the main constituent of pus is dead neutrophils (1 mark)
Dendritic cell	Phagocytic cell with membranous extensions that engulfs pathogens and presents fragments to the cells of the specific immune system; travels into the lymphatic system when activated by a pathogen (1 mark)
Eosinophils	Granulocytes that secrete toxic chemicals that rupture cell walls; located in the bloodstream; especially important in combating parasitic infections (1 mark)
Natural killer cells	Lymphocytes with granules that secrete chemicals that lyse cancer cells and virally infected cells; located throughout the body but in higher concentrations in blood, bone marrow, liver and spleen (1 mark)
Mast cells	Granulocytes that release histamine and are located mainly in the respiratory and gastrointestinal tracts (1 mark)

14 a Inflammation

   b Histamine is released by mast cells. It increases the dilation and permeability of blood vessels including capillaries (1 mark), which influences the swollen, red appearance of the site as fluid and blood engorge the site. (1 mark) This allows for additional proteins and phagocytic cells to migrate to the area and engulf debris and pathogens. Once the cells cannot engulf any more pathogens, they die resulting in pus accumulation. (1 mark) c Complement proteins work in many ways to destroy invading bacteria. They can *(any two of)* (1 mark each):

- bind to the bacteria, acting as a tag for phagocytic cells to facilitate their detection and uptake by these cells (opsonisation)
- induce chemotaxis, creating concentration gradients to attract phagocytes and other white blood cells to the site of infection
- create membrane attach complexes (MAC) that can create pores in the membranes of bacteria, resulting in lysing of the cell.

d Complement is effective against cellular pathogens. Viruses invade host cells so the complement system cannot tag or create pores in 'self' cells.

e Macrophages and dendritic cells will be attracted to the site of infection and engulf pathogens. These cells will then migrate to the lymphatic system. (1 mark) They are antigen-presenting cells that will break down these pathogens and present antigen fragments on their MHC class II markers to helper T cells present in the lymph nodes. In doing this, the helper T cells become activated and the specific immune response is initiated. (1 mark)

15 a *Examples of possible routes include* a cut in the skin, respiratory surfaces, the digestive system, the reproductive system or eyes. (1 mark per route)

b *Suitable examples of chemical barriers and their function include* (1 mark for barrier; 1 mark for function):

- stomach acid, which breaks down bacteria
- lysozymes in tears, which break down bacteria
- complement proteins, which attract immune cells or stimulate phagocytes to become more active by coating their surface and binding.

Some students confused lysosomes with lysozymes.

c *Suitable responses included* vasodilation, or a description of vasodilation or by increasing the permeability of blood vessels.

d Antigen-presenting cells display the antigen (1 mark) on their cell surface to helper T lymphocytes/ helper T cells. (1 mark)

Students needed to describe the initiation of the adaptive immune response by bacterial infection. Responses discussing cytotoxic T cells were therefore incorrect.

16 a Allergen

b Some allergens such as pollen are seasonal and are only released into the environment at different times of the year. Therefore, they are not present in the air all year round.

c i Histamine

ii Histamine increases permeability and dilation of blood vessels. (1 mark) This results in the symptoms exhibited by someone suffering hay fever when the allergen is inhaled. The capillaries in the upper respiratory tract become dilated, bringing more blood to the area, accounting for the redness. The capillaries also become more permeable and the extracellular tissue fluid increases, resulting in both the swelling of the nasal tissues and a runny nose. The response must include a link to the scenario. (1 mark)

d After an initial exposure to the allergen, many IgE antibodies are produced and embedded in the membrane of some mast cells. (1 mark) Upon a second exposure, the IgE antibodies on the mast cells bind to the allergen, cross link and rapidly release histamine, resulting in the symptoms being experienced at a much faster rate. (1 mark)

9780170479431

e *Any two of* (1 mark each):

- Remove known allergens from the person's environment.
- Hang washing inside or use a dryer so allergens do not embed in clothing.
- Avoid touching face and rubbing eyes.
- Close windows in your house and car during high allergen seasons.
- Use filters in vacuum cleaners and air conditioners.
- Take antihistamines to relieve symptoms.

Make sure your responses are specific for the disease in question. Add details so that your response is not just a generic memorised action of histamine.

17 a **i** Skin cell – MHC class I

  ii Macrophage – MHC class II

  iii Red blood cell – none; red blood cells are not nucleated

b MHC class I are on all nucleated cells. These present randomly selected fragments (peptides) from inside the cell cytoplasm for recognition by T cells. This is so that the immune system can survey what is going on inside cells. This method can detect virally infected cells or cancerous cells. (1 mark) MHC class II are only found on antigen-presenting cells such as B cells, macrophages and dendritic cells. This type of marker is used to present extracellular antigens obtained after digesting phagocytosed foreign material such as bacteria. (1 mark) This system is used as a safeguard for the immune system against mounting an immune response to self antigens. (1 mark)

Acquiring immunity

Multiple-choice questions

1 B

Dendritic cells present antigens to helper T cells on MHC class II molecules.

A is incorrect because MHC are not antigens themselves; they present antigens to helper T cells, not cytotoxic T cells. **C** is incorrect because neutrophils do not present entire pathogens, just antigens from them. **D** is incorrect because presentation of antigens occurs in the lymphatic system.

2 B

The lymphatic system is the site where clonal selection and proliferation of B cells occurs.

A is incorrect because clotting factors have to be activated at the site of a wound to seal it. **C** is incorrect because red blood cells carry oxygen around the body and are not involved in the immune response. **D** is incorrect because the initial response to an allergen is triggered at the site of exposure.

3 B

T cells mature in the thymus gland (hence the name T cells) so removal of the thymus gland will reduce the production of mature T cells.

A and **D** are incorrect because this procedure will have no effect on mature B cell production. **C** is incorrect because lymphocytes are produced in the bone marrow, so these will still be active.

4 C

Cytotoxic T cells produce chemicals that can kill virally infected cells and are part of the cell-mediated third line of defence.

A is incorrect because B cells produce antibodies. **B** is incorrect because cytotoxic T cells are part of the specific cell-mediated immune response. **D** is incorrect because cytotoxic T cells only produce cytotoxins. They do not engulf pathogens.

5 C

Plasma B cells must have lots of ribosomes for large amounts of protein translation.

A is incorrect because the size of the nucleus has no bearing on the production of protein. **B** is incorrect because for a cell to produce a lot of proteins they need a lot of energy (lots of mitochondria). **D** is incorrect because a large Golgi apparatus is required for packaging and exporting of the proteins.

6 B

Dendritic cells present antigens to helper T cells.

A is incorrect because helper T cells recognise the antigen. **C** is incorrect because plasma cells are B cells that produce antibodies. **D** is incorrect because mast cells release histamines.

7 B

For the antibody to be effective it has to be a complementary shape to the antigen. Antibody F will be effective against pathogens S, T and U.

A is incorrect because antibody E will be effective against pathogen S only. **C** is incorrect because antibody F will be effective against pathogen U. **D** is incorrect because antibody G will be effective against pathogens R and T.

8 D

Upon exposure to a pathogen for a second time, the immune response is much faster due to memory cells.

A and **B** are incorrect because T cells are not part of the humoral response. **C** is incorrect because plasma B cells produce antibodies and are assisted by helper T cells.

9 C

Active immunity requires the hosts immune system to be activated against a pathogen.

A is incorrect because injecting antibodies is a passive process. **B** is incorrect because monoclonal antibody treatments use injected antibodies to label cancer cells, so they are recognised by the immune system (no production of antibodies by the host). **D** is incorrect because the action of phagocytes is part of the innate immune response.

Short-answer questions

10 a The lymphatic system is a large network of vessels and organs responsible for storage of interstitial fluid (water) to maintain fluid balances within the body (1 mark), and is integral in the storage and transport of immune cells (1 mark).

   b *Any three of* (1 mark each):

   - Vessels – transport of lymph and storage of water and all that is transported or dissolved in it for return back into the bloodstream.
   - Lymph nodes – lymph nodes act as filters. Various immune system cells trap pathogens in the lymph nodes and activate the specific immune response.
   - Spleen – stores immune cells and breaks down red blood cells and platelets.
   - Bone marrow – production of white blood cells.
   - Thymus – T lymphocytes mature in the thymus.
   - Tonsils – contain a lot of white blood cells and, because of their location at the throat and palate, they can stop pathogens entering the body through the mouth or the nose.

   c Primary lymphoid organs – bone marrow and thymus (1 mark)

   Secondary lymphoid organs – lymph nodes, spleen, tonsils and certain tissues in mucous membrane layers (1 mark)

   d No. It does not have a pump like the heart. Lymph flows via pressures created as the body breathes and uses voluntary muscles.

   e The lymphatic system is integral because it is responsible for the production and maturation of immune cells, the storage and transport mechanism for the immune system cells and is the site of antigen recognition and clonal selection and expansion.

11 a Lymph nodes are reservoirs of immune cells that filter the fluid circulating in the lymphatic system around the body. (1 mark) Lymph nodes are placed at the junction of the extremities and the torso so that invaders can be identified before the fluid flows back to the heart, where infection could be life threatening. (1 mark) This is important so that the immune system can quickly pick up any foreign pathogens invading the body to prevent disease. (1 mark)

  b Phagocytes that can engulf and destroy pathogens *or* mast cells release histamines that cause vasodilation. (1 mark for the name and 1 mark for the function)

  c **i** B cells are part of the humoral immune response.

   ii The activation process is called clonal selection and, once activated, the B cell undergoes clonal expansion where it rapidly divides to create effector and memory cells. (1 mark) The effector cells will start to produce antibodies and the memory cells will remain in high numbers within the lymph nodes. (1 mark)

12 a Inflammation has begun to try and rid the body of this viral pathogen. This includes:
  - Fever – the body tries to slow the replication of the pathogen by setting the body's temperature at a higher point and so the rest of the immune system can catch up and fight the infection. (1 mark)
  - Macrophages recognise viral DNA (antigens) as foreign and release cytokines such as interferons and interleukins that have many roles within the immune response. Mast cells release histamines in response to the physical damage to cells when they rupture to release virions to infect other body cells. (1 mark)
  - These chemical messengers increase vasodilation to allow blood, fluid and immune cells to move to the infected area to help fight off an invading pathogen. Additionally, increased permeability of capillaries at the site of infection occurs to facilitate easier movement of immune system cells to the site of infection and also attracts more immune cells to the site of infection. (1 mark)
  - This can be seen by the increase in respiratory and nasal secretions. (1 mark)

  b As the virus is an intracellular pathogen, it will be detected by a cytotoxic T cell when fragments (antigens) of the virus are presented on MHC class I molecules. (1 mark) Once detected, the cytotoxic T cell produces powerful cytotoxins to promote apoptosis of the infected cell. (1 mark) The T_H cells produce cytokines that aid in the proliferation of effector cytotoxic T cells, which continue to destroy infected cells (1 mark), and memory cytotoxic T cells so that a faster immune response will be initiated if the pathogen invades again. (1 mark)

13 a False. Antibodies have very little effect on viruses.

  b True

  c False. Memory cells are stored in the lymphatic system for lifelong immunity. Antibodies are proteins, so only have a relatively short life span.

  d True

  e False. Both B and T cells undergo clonal selection and clonal expansion.

  f True

  g True

  h False. Cytotoxic T cells produce powerful substances that kill virus-infected cells.

14 a B cells

  b   (3 marks)

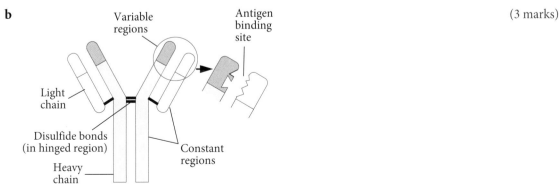

SOLUTIONS

When drawing an antibody, make sure you know the orientation of the light and heavy chains and where they are bonded together.

   c IgG

   d IgM

   e Agglutination – Pathogens become trapped in a network of antibodies, immobilising them and making them more susceptible to destruction. (1 mark)

   Opsonisation – Bound antibodies 'tag' pathogens for destruction by phagocytic cells, making identification of the pathogen easier. (1 mark)

   Neutralisation – Antibodies can bind to toxins, preventing them from binding to target cells. (1 mark)

   Complement activation – Bound antibodies activate the cascade of complement proteins. (1 mark)

15 a Passive artificial

   b Passive natural

   c Active natural

   d Active artificial

For the response to be 'active', the person's immune system must be activated to produce memory cells. For the response to be 'natural', it must be a naturally occurring situation or process.

Disease challenges and strategies

Multiple-choice questions

1 C

Zoonosis is the term for a disease that can be passed from animals to humans through many different routes.

A is incorrect because it is nonsense. **B** is incorrect because the disease is passed from animals to humans. **D** is incorrect because a zoonosis is cross-species infection.

2 D

All of the options may be causes for a disease to re-emerge in a community except travellers visiting developing countries. In this scenario, only a few (presumably healthy) people are travelling to a place with the disease.

3 D

The most effective of these options is isolating infected individuals, because it is an airborne pathogen.

A and **B** are incorrect because washing hands and clean needle programs will work only a small amount. **C** is incorrect because vaccinating infected people will have no impact. Vaccinations are only effective in non-infected people.

4 B

Polio and smallpox have similar infection rates of 5 to 7 people infected from one case.

A is incorrect because the most contagious disease appears to be measles or whooping cough, as one person can infect a larger number of people. **C** is incorrect because death rates are not mentioned. **D** is incorrect because the faecal–oral route is the mode of transmission for only one disease that does not infect a large number of people from one case. Airborne transmission is the most effective mode of transmission of diseases in the table.

5 D

Polio is endemic to Pakistan because there is a consistent small number of cases each year.

A is incorrect because an epidemic is a large number of infections in a community or area. **B** is incorrect because a pandemic is a worldwide infection. **C** is incorrect because an outbreak is a higher-than-normal infection rate within a small area or town.

6 A

The parasitic vector is a mosquito as it carries the virus from host to host.

B and **C** are incorrect because the parasite is carried by the mosquito and passed on to humans. **D** is incorrect because malaria is treatable and preventable because malaria is an infectious disease (mosquito vector) but not a contagious disease.

7 D

Preventing the mosquitos from biting humans is the most effective prevention measure.

A is incorrect because nearly half of the world's population live in malaria-prone areas. **B** is incorrect because eliminating mosquitos would have devastating environmental impacts. **C** is incorrect because increasing spending on treatments does not reduce the incidence of disease.

8 C

Often citizens of developing nations have limited health care available.

A is incorrect because the parasite does not change its surface proteins like a virus can. *Plasmodium* is an intracellular parasite so, if it keeps changing its surface antigens, there is no exposure to the immune system because it is inside the red blood cell. **B** is incorrect because mosquitos do not tend to travel long distances. **D** is incorrect because mosquitos only live in warm environments, so travellers returning to colder climates would not have the correct species of mosquito for the disease to spread.

9 A

Just because the family members do not display symptoms does not mean they are not carrying the virus and symptoms may appear after a few days.

B is incorrect because a response is required to prevent the transmission of disease. **C** is incorrect because antibiotics will have no effect on a viral infection. **D** is incorrect because returning the family to Afghanistan is an overreaction.

10 A

The graph shows the vaccination on day one with the first peak in antibodies, and then a second exposure to the same strain of influenza in the second peak that begins on day 22. Memory B cells will have been formed from the vaccination that results in a much quicker and larger response to the second exposure.

B is incorrect because B plasma cells would have been most numerous around day 34–36, as this is where the highest level of antibodies is recorded. **C** is incorrect because herd immunity involves a community, not just one person. **D** is incorrect because vaccination occurred on day 1.

11 B

Monoclonal antibodies are proteins produced from the same clone of a cell and are specific to the same antigen.

A is incorrect because monoclonal antibodies are proteins. **C** is incorrect because monoclonal antibodies will bind to receptors on the outside of the cell. **D** is incorrect because monoclonal antibodies only treat one type of cell.

12 B

Monoclonal antibodies make the cancer cells more visible to the host's immune system.

A is incorrect because they do not cross the cell membrane, so do not bind to DNA. **C** is incorrect because they are not antigens, so do not trigger the patient's own immune system. **D** is incorrect because they do not lyse cells.

13 C

In order to detect an infectious disease, one process is via the use of an antigen–antibody complexes. Monoclonal antibodies can be used to detect pathogens within a patient's blood.

A is incorrect because the detection of an infectious disease is required, not the detection of self cells. **B** is incorrect because an antigen needs to be detected by an antibody. Antigens do not recognise antigens. **D** is incorrect because pathogens do not produce antibodies.

Short-answer questions

14 Indigenous communities were isolated for long periods of time so did not develop immunity to many common pathogens found in European communities. (1 mark) As a result, when Europeans settled in Australia, they brought with them many diseases that indigenous communities had very little exposure to. Many succumbed to these diseases due to lack of any type of exposure and therefore immunity to these pathogens. (1 mark)

15 a Eukaryotic single-celled organism with no nucleus (1 mark); circular DNA and no membrane-bound organelles (1 mark)

b Airborne. (1 mark) This disease causes a cough, which can disperse the bacteria into the air. It also infects the lungs so must be breathed in. (1 mark)

c The raised red patch is an immune response to tuberculin. If someone has been exposed to tuberculin in the past, they will have immune cell memory and a reaction will occur due to the second exposure to it.

d Reduced vaccination rates within Australia (1 mark) and increased immigration of people from overseas countries where vaccination programs are not implemented or disease is endemic (1 mark)

> Use the information within the stem of this question to answer the multiple parts and keep referring back to it as evidence for your statements.

16 a *Factors that could lead to transfer between hosts include* (1 mark for one factor):

- increased population coming into contact with wild animal habitat
- urbanisation of habitats in proximity to wild animals increases the chance of direct transfer
- transportation of wild animals
- lack of education about transmission of disease.

> Transmission is not via food or water.

b Specific pathogens have specific methods of transfer. (1 mark) They may require specific methods of control such as a particular antiviral drug. (1 mark) Incorrect identification can lead to continued infections and spread. (1 mark) *or*

Identify pathogen and isolate antigens. (1 mark) This would enable the production of a vaccine or produce a drug with a complementary shape. (1 mark) If the majority of the population was vaccinated or treated with this rational drug, this would greatly reduce the transmission of the disease. (1 mark)

c The shape of the Zika antigen and the antigen of the other viruses is similar, enabling them all to bind on to the same antibody. (1 mark) The viruses bind to the antigen-binding site of the antibody. (1 mark)

d *Three different approaches could include* (1 mark each):

- inspect and fumigate cargo entering new countries
- quarantine of people who come from countries in which the disease occurs
- control mosquito breeding places, use of mosquito nets or fly screens on windows
- education campaign about suitable clothing, use of insect repellent.

17 a Cytotoxic T cells recognise antigens presented on MHC class I molecules. (1 mark) They then produce cytotoxic substances to destroy virally infected cells. (1 mark)

b A live attenuated virus is a live virus that has been changed or weakened so that it does not cause symptomatic disease. (1 mark) It is used because it creates a strong immune response due to the virus remaining mostly intact, so the immune system can recognise and mount a response to viral antigens. (1 mark)

c i

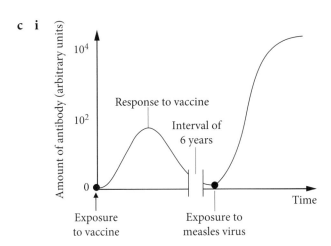

ii The second exposure to the measles virus results in a faster and larger antibody response due to the presence of memory cells in the immune system.

d While both mumps and measles are paramyxoviruses, they have different antigens on their surface. (1 mark) The immune system creates antibodies that are very specific, so there would be no protection against mumps from this vaccination due to the specificity of the antibodies produced by the body. (1 mark)

18 a i 1958

ii Observed trends are, for example, when the number of cases increases there is an increase in the number of deaths (1 mark) and the numbers of cases and deaths both decline (1 mark).

You must include a comparative statement in your answer to receive full marks.

b Herd immunity is where most of the community is immune (1 mark) and this helps to protect, for example, babies or those few individuals who cannot be vaccinated, or due to the reduced number of infected individuals there are fewer hosts to pass the disease to others. (1 mark)

19 a A disease in which the body produces antibodies or otherwise mounts an immune attack against its own cells

b The ability of the body to distinguish self from non-self

c i They lower the activity of the immune cells, thus limiting the attack on self tissue.

ii The body will be less likely to be able to fight pathogens such as bacteria and viruses (1 mark) and less able to destroy damaged self cells such as cancer cells. (1 mark)

d Monoclonal antibodies are very specific and are used to inactivate molecules involved in the immune response, such as cytokines that cause the inflammation. (1 mark) Therefore, they do not result in a general immunosuppression, and hence the patient is less susceptible to secondary infections. (1 mark)

CHAPTER 4, UNIT 4 AREA OF STUDY 2

Genetic changes in a population over time

Multiple-choice questions

1 B

A gene pool is the total sum of a population's genetic material at a given time.

A is incorrect because this is just describing a habitat. **C** is incorrect because a gene pool is associated with one species only. **D** is incorrect because a gene pool involves all members of the population.

2 A

A nucleotide has been inserted into the sequence, causing a frameshift mutation.

B is incorrect because an inversion involves the switching of two bases in the sequence, resulting in a sequence of the same length. **C** is incorrect because a deletion involves the deletion of one or more bases, resulting in a shorter fragment. **D** is incorrect because translocation involves large parts of a chromosome, not single nucleotides.

3 A

A missense mutation occurs when a single nucleotide change results in a codon that codes for a different amino acid to the original one.

B is incorrect because a nonsense mutation causes the amino acid chain to stop. **C** is incorrect because a frameshift mutation results in the amino acids being different from the mutation onwards. **D** is incorrect because a silent mutation causes no change to the amino acid sequence.

4 C

A mutation is not necessarily beneficial. It could be detrimental or silent.

A, **B** and **D** are incorrect because they are all true statements.

5 A

Genetic drift is when allele frequencies change due to chance. This is because the bears have been isolated and the population will have responded to environmental pressures over multiple generations.

B is incorrect because the question does not describe separation into two different species. **C** is incorrect because this is just one trait in one species, not two species evolving similar traits separately. **D** is incorrect because it is just one trait and the bears can still produce fertile offspring.

6 C

DDT will kill all susceptible mosquitos and leave those resistant to it. Those mosquitos left will breed, passing on their genes to the next generations, resulting in populations that include more and more mosquitos that are resistant to the insecticide.

A is incorrect because the insecticide does not cause mutation in insects. **B** is incorrect because this insecticide kills sensitive mosquitos, so a tolerance does not develop. **D** is incorrect because the question only documents one insecticide.

7 A

Increasing genetic isolation means that the soapberry bugs are unable to interbreed. Host plant fruiting seasons will result in this.

B, **C** and **D** are incorrect because they all still allow interbreeding of populations.

8 C

Adding up the alleles 'R' and 'r' in Generation 1 results in them being equal.

A is incorrect because genetic diversity is decreasing in this population. **B** is incorrect because the genotype 'rr' had a selective advantage. **D** is incorrect because there is no evidence of new alleles.

9 D

Reduction in numbers results in a larger number of individuals having the same alleles, leaving them vulnerable to disease due to a decrease in variation within the population.

A is incorrect because there is a reduction in numbers influencing alleles, not environmental pressures. **B** is incorrect because it is nonsense. **C** is incorrect because this is a consequence to hunters, not to cheetahs.

10 B

This population has less genetic variation than the Arizona populations and is an example of the founder effect.

A is incorrect because the gene pools on Tiburon Island and Arizona cannot possibly be identical. Only 20 individuals were taken to the island; not enough to include all variations of all alleles. **C** is incorrect because the mutation rate will not be any higher on the island than anywhere else. **D** is incorrect because the current population will show decreased genetic variation compared to the Arizona populations.

11 B

Human intervention is shaping the allele frequency within a population.

A and **C** are incorrect because this is not a naturally occurring process, and can occur within a relatively short period of time over only a few generations. **D** is incorrect because this is not a sharp reduction in the size of a population.

12 A

The antibiotics kill the most sensitive bacteria first. The resistance of bacteria to antibiotics is due to natural selection, where the less sensitive bacteria survive and reproduce.

B is incorrect because antibiotics do not treat viral infections. **C** is incorrect because antibiotics don't cause mutations in bacteria. **D** is incorrect because human cells must not be affected by the antibiotics.

13 C

If bacteria become resistant to antibiotics, patients will not be able to be treated, resulting in uncontrolled infections.

A is incorrect because hospital stays become longer, and intravenous drugs are required. **B** is incorrect because antivirals do not treat bacterial infections. **D** is incorrect because tolerance is a term used for drug use, not treatments for infection.

Short-answer questions

14 a   Genetic drift – when allele frequencies change over generations due to chance

   b   Environmental selection pressures may have influenced the survival of the larger bugs over the smaller ones. (1 mark) *Any suitable example of a selection pressure would be accepted here. For example:* larger bugs may have been able to overpower smaller ones for mates; larger bugs consume more of the available food, therefore live longer etc. (1 mark)

   c   Gene flow is the process of alleles moving from one population to another; for example, when two populations of rhinoceros beetles are able to interbreed. (1 mark) Genetic drift is the alteration of allele frequency in a gene pool; for example, the beetles become larger over subsequent generations. (1 mark)

15 a   UUA GCC AUC

   b   **i**   The mutation is a substitution, where one nucleotide (T) is substituted for another (C).

      ii   UUG GCC AUC

      iii   The triplet UUA (original strand) and UUG (Mutation 1) both code for the amino acid Leu (leucine). (1 mark) There would be no change to the peptide because this results in a silent mutation. (1 mark)

   c   Mutation 2 – This is an insertion mutation because one extra nucleotide is inserted into the DNA strand (1 mark) resulting in the reading frame for the codons being moved forward by one nucleotide. The result will be that all amino acids past the mutation point will be different. (1 mark)

16 a   **i**   When a new population is established from only a small number of individuals, resulting in a loss of genetic variation within the new population.

      ii   The 200 founding individuals did not have the same level of genetic variation as the original population. (1 mark) As a result, the frequency of some alleles has increased as the population increased. (1 mark)

   b   Due to the reproductive isolation within the Amish community and low levels of genetic variation, the occurrence of this autosomal recessive trait has been passed on to future generations and increased with the increase in population size from 200 founders to more than 7000 people.

   c   This is an autosomal recessive trait, so much of the community could be carriers but unaffected by the disease.

   d   This will result in a further increase in the allele frequency within the population.

17 a Natural variation exists in a population through mutations (1 mark) that may have created new alleles or by different allele combinations in sexual reproduction. (1 mark) Changes in chromosome number may change phenotype. (1 mark)

   b The lower genetic diversity was due to the founder effect. (1 mark) The two individuals will not have all the variations of alleles as seen in the original population, (1 mark) the descendants of these two individuals will not be able to inherit other alleles and there will be less variation in phenotypes of the lizards. (1 mark)

   c The decrease may have occurred due to different selection pressures on the smaller islands. (1 mark) Over this time period, those lizards with shorter hind legs are at an advantage, and are more likely to survive, reproduce and pass the trait to their offspring. (1 mark)

> Some students incorrectly stated that the change was due to a lack of use of the hind legs or the lizards' needs driving the change.

18 a Mutation

   b They breed multiple times a year, producing multiple offspring.

   c In any population of rabbits, genetic variation exists (1 mark), which may give an advantage to some rabbits when the environment changes and temperature increases (1 mark). Only those rabbits that can tolerate hotter climates (such as those with alleles for thinner coats) will survive long enough to reproduce and pass on their traits to their offspring. (1 mark) This will result in an increase in thinner-coated rabbits over subsequent generations, increasing the allele frequency in the population. (1 mark)

19 a   **i** Natural selection: the process where particular traits are better suited to the environment than others, and as a result the 'fittest' will survive and reproduce, influencing the characteristics of the population (allele frequencies). (1 mark) Example *(any relevant environmental example accepted)*: thicker coated polar bears will survive longer in extended winters. (1 mark)

   ii Selective breeding: when humans intervene in breeding processes and choose which animals will reproduce. (1 mark) Example: corn crops that contain only yellow kernels will be harvested for the following year's seed. (1 mark)

   iii Genetic engineering: the process of using recombinant DNA technology to alter the genetic makeup of an organism. (1 mark) Example: including an insect resistance gene into cotton seeds to stop insects from destroying the crops. (1 mark)

   b Genetically modified organisms are organisms that have had their DNA altered in some way. This could include 'turning on/off' genes or the introduction of new genes into the genome. (1 mark) Transgenic organisms are created when the DNA from one species is inserted into a different species. They are a type of genetically modified organism. (1 mark)

   c Advantages *(two of)* (1 mark each): more productive crops; animals with larger muscle mass; animals with no horns (reduces risk to farmers); more tolerant crops to different soil types etc. Disadvantages *(two of)* (1 mark each): loss of species variation, leaving it vulnerable to disease; can create desired physical traits that are not healthy for the actual organism.

20 a The innate immune response

   b   **i** Antigenic variation is the mechanism pathogens use to alter the proteins or carbohydrates on its surface to avoid a host immune response.

   ii Attempts to develop a vaccine have been unsuccessful due to the specificity of the host immune response. The host will develop immune memory cells to specific antigens (1 mark) but, because the protozoa changes its surface biomolecules, the host immune system does not recognise the pathogen. Every time this happens the host body must create new specific antibodies and memory to the changed surface antigens. (1 mark) Thus, the protozoa can evade the immune response of the host, rendering a vaccine unsuccessful. (1 mark)

   iii Control the tsetse fly population

   c The pathogen is able to replicate multiple times before the host expires.

d Antibiotics are only effective against gonorrhoea (bacteria) (1 mark) as they interrupt binary fission or disrupt the cell membrane because they are single-celled organisms. (1 mark)

e The most susceptible bacteria will be killed first. Antibiotics must be taken to maintain concentration within the body so that all bacteria are killed. (1 mark) If the full course of antibiotics is not taken, the most resistant bacteria remain, and this will contribute towards the development of antibiotic-resistant bacteria. (1 mark)

21 a H1N1 identifies the type of influenza virus by the proteins on the outer shell. (1 mark) The 'H' (haemagglutinin) and the 'N' (neuraminidases) are both proteins that are found on the outer shell or envelope of the virus. (1 mark) Different viruses have different haemagglutinin and neuraminidase proteins.

b Antigenic drift is the accumulation of minor genetic mutations that occur over time. (1 mark) Antigenic shift is a major change that occurs when two or more different strains combine to form a new subtype that displays a mixture of surface antigens from both subtypes. (1 mark)

c A pandemic is a disease outbreak that spreads worldwide, or over a very wide area, crossing international boundaries and usually affecting a large number of people. Swine flu breached the borders of Mexico and spread throughout the world.

d This is not a sensible precaution. The name 'swine flu' originated because the virus was a mutated form of a pig influenza virus. It is not found in pig muscle tissues; it is a respiratory virus.

e It is a respiratory virus, so viral loads will be higher in the lungs; therefore, it would be more effective to directly treat the cells infected. (1 mark) If the drug was ingested, the acidic chemical environment of the gastrointestinal tract may affect the drug and reduce its effectiveness. (1 mark) *or* Inhalation will be a much faster treatment because it comes in to contact with the infected cells (1 mark), rather than having to be absorbed by the gastrointestinal tract then travel through the bloodstream to get to the infection site. (1 mark)

f Influenza viruses change their surface proteins; therefore, the immune system needs to be continually challenged with different antigens to maintain a variety of specific memory cells.

Changes in species over time

Multiple-choice questions

1 C

The picture shows how a species changes over time, so it demonstrates transitional fossils.

A is incorrect because these are not in abundance and not used to determine the time frame for other fossils. **B** is incorrect because these are bones from an organism and not evidence of life. **D** is incorrect because the question gives no information as to whether or not these were complete organisms.

2 D

The fossil depths also indicate the age of the fossils, so *G. obliqua* is the oldest because it was found deeper in the coal deposit.

A is incorrect because they are all of the same genus. **B** is incorrect because *G. obliqua* is deeper in the strata so *G. clarkeana* cannot be an evolutionary ancestor. **C** is incorrect because carbon dating is only used for fossils younger than 60 000 years. These fossils are 245–290 million years old.

3 B

The fossil has undergone three half-lives resulting in one quarter carbon-14 and three quarters nitrogen-14.

A is incorrect because there is no information about where the fossil was found. **C** is incorrect because carbon dating is effective in fossils less than 60 000 years old. **D** is incorrect because carbon decays, resulting in a decrease in carbon and an increase in nitrogen.

4 D

A is incorrect because this is just reproduction. **B** is incorrect because this is just a species exploiting an environmental niche. **C** is incorrect because, for speciation to occur, fertile offspring must be produced.

5 A

> Allopatric speciation occurs when two populations of the same species become isolated from each other due to geographic changes.
>
> **B** is incorrect because this is just environmental changes. **C** is incorrect because this is the beginning of divergent evolution. **D** is incorrect because this is environmental change that will be detrimental to the species as a whole.

6 B

> A prezygotic reproductive barrier is a mechanism that prevents fertilisation from occurring. The different flowering seasons would prevent fertilisation from even occurring.
>
> **A**, **C** and **D** are incorrect because the barrier must occur before fertilisation.

Short-answer questions

7 a Strata (1 mark); They are formed as different layers of sediment get laid down over time. (1 mark)

    b No. The surrounding sediments are different, so they would have been alive at different times.

    c

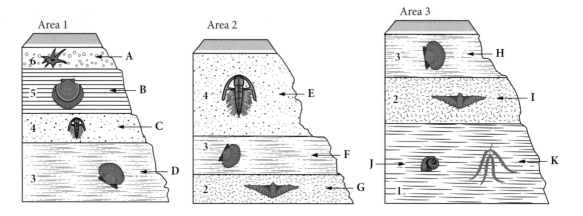

> 1 mark for correct direction (1 being oldest) and 1 mark for correct layers

    d An index fossil is any animal or plant preserved in the rock record of Earth that is found in abundance over a particular span of geologic time or environment.

    e No. Only organisms with hard body parts are fossilised, so it is an incomplete record. (1 mark) Organisms with soft bodies cannot be fossilised. (1 mark)

8 a Over time, mutations occur in the DNA of the common ancestor of these plants, giving them a selection advantage within their local environments. (1 mark) The advantage could have been *(at least one of these, or any other relevant example)* an abundance of small insects pollinating plants with smaller flowers in the south east, larger birds only pollinating larger flowers in the north, or temperature differences selecting for those individuals that could withstand higher temperatures in the north. (1 mark) Due to this selection advantage, over time only the fittest offspring were able to reproduce (1 mark), resulting in enough changes that the pollen from one plant (*E. viminalis*) will not produce viable offspring in the other (*E. miniata*). (1 mark)

    b No. *(Two of the points below)* (1 mark each)

    E. miniate is pollinated by birds whereas *E. viminalis* is pollinated by insects.

    E. miniate is pollinated in the winter months whereas *E. viminalis* is pollinated in the summer months.

    E. miniate is located in south-east Australia whereas *E. viminalis* is located in the north resulting in it being too far for pollinators to travel.

> It is not enough to simply state 'No' as the answer without giving two pieces of evidence. The pieces of evidence gain the marks.

9780170479431

c The bottleneck effect is when there is a sharp reduction in the size of a population due to environmental or human events. (1 mark) The remaining trees may not be a representative sample of the whole genetic variation that once occurred in the forest. (1 mark) This would result in decreased genetic variation within the remaining plants. (1 mark)

d No, carbon dating is only used for fossils that are less than 60 000 years old. These fruits are 52 million years old.

e They would have fallen from the tree and been immediately covered in sediment or ash, creating an environment lacking in oxygen. (1 mark) Over time, more and more sediment layers would settle on top. Mineralisation occurs where minerals from the surrounding environment seep in and replace the organic material in the fossil. (1 mark) The hard outer shell of the fruits is more conducive to fossilisation than softer parts of the plant.

9 a He hypothesised that they had all come from a common ancestor from the mainland.

  b Size, beak shape and claw size

  c Environmental conditions and abundance of food sources such as insects or nuts/seeds

  d If the finches primarily ate insects, they had longer beaks and sharper claws, whereas if their food source was nuts and seeds, they had wider, more powerful beaks for cracking them open. (1 mark) The birds that were most suited to the environment were able to thrive and reproduce, resulting in a change in allele frequency within the different populations of finches on the different islands, and they therefore no longer resembled the parent population on the mainland. (1 mark)

  e The theory of evolution by natural selection

10 a The hypothesis is supported because the average skull measurements in Population B are different from those in Population A. For example, in Position 1 the difference is 20 mm. (1 mark) The differences in measurements in the three other species are smaller. (1 mark)

> Students are required to refer to the data provided in the table to justify their answer.

  b *One of, including explanation* (2 marks):
- Comparison of DNA sequences: evidence to support their hypothesis would be differences in the nucleotide sequences within the DNA molecules or evidence from DNA hybridisation studies
- Comparison of mitochondrial DNA: evidence to support their hypothesis would be differences in the nucleotide sequences within the DNA molecules
- Comparative genomics (comparison of whole genome sequences): evidence to support their hypothesis would be differences within the gene sequences of the two populations

  c **i** A geographical barrier

> Incorrect answers relate to food availability.

    ii A geographical barrier prevents the two populations from breeding together so there is no gene flow (1 mark), and different selective pressures may act on the populations. (1 mark)

> Most students recognise that gene flow does not occur but miss the second point.

11 a Soil preferences caused a shift in the flowering of these two related species. Because they flower at completely different times when growing on their respective soil types (1 mark), enough reproductive isolation was introduced to disrupt the random fertilisation of these wind-pollinated palms. (1 mark) As soon as such reproductive biases are introduced, speciation can and will occur.

  b Sympatric speciation

Determining the relatedness of species

Multiple-choice questions

1   A

Homologous structures are similar due to a common ancestry, but may have evolved for different uses.

B is incorrect because analogy is when structures have a similarity of function and superficial resemblance of structures but have different evolutionary origins. **C** is incorrect because convergence is when two or more species develop independently but result in similar structures. **D** is incorrect because an arm and a flipper are not the same.

2   B

Due to a common ancestor at some point, they both contain the similar genes that are active during the embryonic stages but are switched off sometime during development.

A is incorrect because the common ancestor is not recent. **C** is incorrect because so many animals share these characteristics that it is unlikely to be the result of convergent evolution. **D** is incorrect because modern fish do not have species evolved from them.

3   B

These structures have evolved for the same purpose, but a different ancestor indicates that this is an example of convergent evolution, because the vultures evolve similar traits as a result of having to adapt to similar environments.

A is incorrect because having different ancestors eliminates this response. **C** is incorrect because genetic drift is a change in allele frequencies over time. This does not answer the question. **D** is incorrect because migration is the movement of animals and does not explain ancestry in this context.

4   B

The node between Species A and Species B is farthest to the right, indicating a most recent common ancestor.

A is incorrect because the phylogenetic tree shows Species A and B closest together (indicating they would have the most similar DNA sequences). **C** is incorrect because Species C shares a common ancestor with A and B but is not the common ancestor itself. **D** is incorrect because all four species share a common ancestor but not mitochondrial DNA, as this is passed on only from the maternal parent.

5   C

Using the phylogenetic tree, follow the line for *O. latipes* back to the first node showing the split between Beloniformes and Cichliformes. These are the most closely related species.

A, **B** and **D** are incorrect because they are further away on the phylogenetic tree.

6   D

Use the timeline underneath the diagram to identify a timeline for the evolution of this species. The node for these two species is at 25 million years ago.

A is incorrect because cichlids diverged to form three distinct species only 5–10 million years ago. They had a common ancestor 100 million years ago. **B** is incorrect because *C. chanos* was the first species to diverge from the most distant common ancestor. **C** is incorrect because Gasterosteiformes, Beloniformes and Cichliformes do share the common ancestor Acanthopterygii.

Short-answer questions

7   **a**   Divergent evolution

    b   35 million years ago

    c   Marsupial mole and bilby/bandicoots

9780170479431

d i

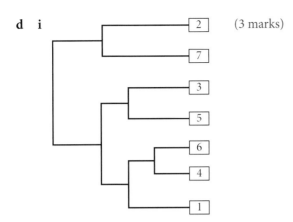

(3 marks)

ii The coppery ringtail and musky rat-kangaroo have similar claw structures, so are the most closely related compared to the other hind feet pictured. (1 mark) Black dorcopsis and the black-footed rock wallaby have thinner foot structures and similar features but also have similarities to Bennett's tree-kangaroo, grizzled tree-kangaroo and Lumholtz's tree-kangaroo. This reflects that they share a common ancestor. (1 mark) Bennett's tree-kangaroo, grizzled tree-kangaroo and Lumholtz's tree-kangaroo have similar morphology; however, their claw lengths are slightly different. The species Bennett's tree-kangaroo and grizzled tree-kangaroo seem to be more closely related because their claws are larger than that of Lumholtz's tree-kangaroo. (1 mark)

8 a DNA is heated to break the bonds/separate the strands (1 mark) and single strands from both species are mixed together and cooled. (1 mark) The resulting hybrid DNA is then reheated and the temperature at which the DNA becomes single stranded is noted. (1 mark)

b When heated for the second time, the fossil DNA mixed with the crane DNA will require more heat (higher temperature) before it will separate due to the molecular homology. (1 mark) The temperature at which the DNA becomes single stranded indicates the degree of complementary base pairing (1 mark) and, the higher the temperature, the more closely related the two species are. (1 mark)

This is not PCR. The procedure is very different.

9 a The length of the line suggests that *Pakicetus* is extinct.

b *Two of* (3 marks):

Aquatic adaption/advantage	Terrestrial adaption/advantage
Heavy bones to enable organism to sink/stay submerged	Flat feet/hooves to walk on land
Paddle-like limbs/webbed feet to assist swimming	Four walking legs to walk on land
Nostrils/eyes at top of head for surface breathing/vision	Short legs for walking in mud

Students were required to give a feature of the transitional fossil that would be advantageous in an aquatic environment and another feature that would be advantageous in a terrestrial environment. They were also required to provide an advantage of each given feature. The feature needed to be feasible in a fossil. Presence of lungs or gills is not feasible, as these would not fossilise.

c i *Examples of possible responses include* (1 mark each): The hippopotamus and the killer whale both have a relatively recent mammal ancestor 55 million years ago or diverged under different selection pressures and adapted to different environments. *or*

The bronze whaler and the blacktip reef shark both have a shark ancestor 23 million years ago *or* diverged under different selection pressures and adapted to different environments.

ii *Examples of possible responses include* (2 marks): Humpback whale and whale shark: both adapted to ocean swimming with big, streamlined bodies and tails. They are both filter feeders, straining plankton through sieve-like structures in the mouth. *or* Killer whale and bronze whaler: both have a dorsal fin for efficient swimming.

Some students confused divergent and convergent evolution, either by providing examples or explanations of the other type of evolution or by answering the opposite for each type of evolution.

d Evidence could be gained by interbreeding the burrunan dolphin and the common bottlenose dolphin to see if they produce fertile offspring. (1 mark) They could also use DNA hybridisation, using DNA samples from both species to see if there are many complementary DNA sequences. (1 mark)

e *Possible outcomes could be that (two of)* (1 mark each):
- The burrunan dolphin evolves into a species separate from that in the Gippsland Lakes as there are different selective pressures acting on the populations.
- The burrunan dolphin becomes extinct due to human impact pressure/overfishing/disturbance/pollution.
- If a conservation group looks after the dolphin groups, they may increase in number.

Human change over time

Multiple-choice solutions

1 D

A and C are incorrect because the tree shrew and birds are not closely related to hominoids. **B** is incorrect because gorillas are hominoids.

2 B

All primates have flat nails instead of claws and dextrous hands.

A is incorrect because primates do not have tails. **C** is incorrect because primates do not always have arms longer than legs and even-sized teeth. **D** is incorrect because not all primates have a bipedal stance.

3 C

Hominins differed from apes by being bipedal.

A is incorrect because, while cranial capacity is increasing, hominins do not have significantly larger brains. **B** is incorrect because they did not have less body hair. **D** is incorrect because all primates live in family groups.

4 D

A is incorrect because *Homo neanderthalensis* have slightly larger brain sizes. **B** is incorrect because *Homo neanderthalensis* inhabited northern Africa and Europe. **C** is incorrect because *Homo neanderthalensis* lived at the same time as modern humans; they are not their ancestors.

5 C

A is incorrect because *Homo neaderthalensis* is extinct. **B** is incorrect because *Australopithecus* are classified as hominins. **D** is incorrect because all hominins are hominoids.

6 A

Skull 3 is more ape-like, followed by Skull 1 where the face is flatter and cranial capacity increases, followed by Skull 2, which looks more like a modern human skull.

7 D

Look for at least one contradictory statement that does not support the question. *Homo* species do not have more-opposable big toes, the jaw size is decreasing, and zygomatic arch is decreasing in size.

A is incorrect because zygomatic arch is decreasing in size. **B** is incorrect because *Homo* species do not have more-opposable big toes. **C** is incorrect because the jaw size is decreasing.

8 D

As there is only a small percentage of Denisovan DNA in modern humans, they are not descended from them; rather, there must have been interbreeding to produce viable offspring.

A is incorrect because the technology used is improving and it is unlikely that the DNA sequencing was incomplete. **B** is incorrect because Denisovans migrated from Eurasia, not Africa. **C** is incorrect because there is only a small percentage of Denisovan DNA in modern humans; therefore, they are not descended from them.

9 B

As *H. sapiens* DNA was found in the 100 000-year-old fossils, they must have come in to contact with Neanderthals prior to 65 000 years ago.

A, **C** and **D** are incorrect because modern Africans do not contain Neanderthal DNA.

10 A

As the fossil record is incomplete, it is likely that scientists haven't found the early form of *Homo* species yet.

B is incorrect because, while there is some margin for error, 500 000 years is too big of a gap to be attributed to error. **C** is incorrect because, while it is true, it does not answer the question asked. **D** is incorrect because Australopithecines were able to grasp items.

11 D

Mitochondrial DNA is passed on from mother to child, so is only useful for chimpanzees and humans.

A, **B** and **C** are incorrect because the organisms do not reproduce sexually, and cannot have DNA passed from the maternal parent.

12 C

A is incorrect because the proteome of species was not investigated. **B** and **D** are incorrect because the comparison between the different species of monkeys was not investigated.

13 A

As there are two amino acid differences, the DNA sequence must differ by at least two nucleotides.

B is incorrect because, due to the DNA code being redundant, we do not know whether the DNA sequence that codes for this protein is identical. **C** is incorrect because there are fewer differences between humans and gorillas, indicating that they share a more recent common ancestor. **D** is incorrect because gorillas and chimpanzees are not investigated.

Short-answer questions

14 (1 mark for each word or category placed in the right location on the diagram)

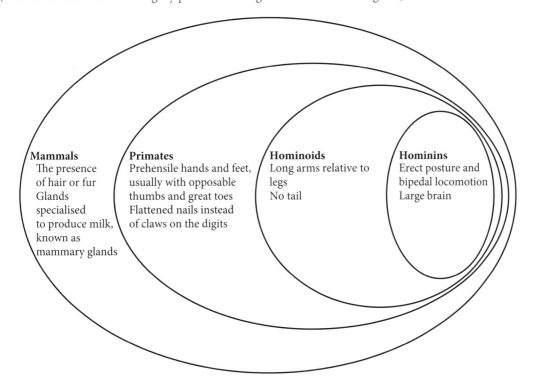

Mammals	**Primates**	**Hominoids**	**Hominins**
The presence of hair or fur Glands specialised to produce milk, known as mammary glands	Prehensile hands and feet, usually with opposable thumbs and great toes Flattened nails instead of claws on the digits	Long arms relative to legs No tail	Erect posture and bipedal locomotion Large brain

15 a *Two of* (1 mark each)*:* Prehensile hands and feet, opposable thumbs, 3D colour vision, social living

b i *Two of* (1 mark each)*:* Bipedalism, erect posture, larger brains, behavioural characteristics such as specialised tool use

ii *Example answer:* Bipedalism is an advantage because it frees the hands for carrying tools and infants, improves the ability to cool-off, allows the organism to see over tall grasses and allows the travelling of long distances.

iii *Homo erectus*

c Brain size gradually increased over time as human species evolved from *Australopithecus* through to *Homo* species, and finally to *Homo sapiens*.

16 a There has been an increase in the quantity of research conducted, as well as technological advances.

b DNA analysis (1 mark) and morphological characteristics (1 mark)

c Modern humans, bonobos and chimpanzees share a recent common ancestor.

d That it is incomplete and is constantly changing when new fossil evidence is discovered

e It is unlikely. With the addition of DNA technologies, relatedness of species can be determined much more closely than morphological evidence alone. (1 mark) It is more likely to enhance current understandings and create a 'bushier' family tree rather than completely change it. (1 mark)

f *Ramapithecus* would be moved from the humans line and branched off the orangutan line.

17 a i Interbreeding occurred between *H. neanderthalensis* and ancestors of (present-day European, East Asian and Australian Aboriginal) *H. sapiens*. (1 mark) DNA was passed from one generation to the next. (1 mark)

> Many students correctly identified that gene flow occurred. Others, however, incorrectly identified this as genetic drift.

ii They are not separate species.

b Out of Africa (1 mark) – Humans first evolved in Africa where there were no Neanderthal populations because there is no Neanderthal DNA in populations in Africa, (1 mark) and populations of humans moved out Africa and encountered other *Homo* species. This is supported by the presence of *Homo neanderthalensis* DNA in all modern humans except African populations. (1 mark)

c Route: Africa → Middle East → East Asia → Australia (1 mark)

or

Europe → East Asia → Australia (1 mark)

Timing: Migration occurred sometime less than 80 000 years ago. (1 mark)

18 a Different palaeontologists may make different interpretations of the same data.

b i *Either: Australopithecus afarensis* evolved about three million years ago. *or Homo heidelbergensis* evolved from *Homo ergaster* (about one million years ago).

ii *Any one of: Homo erectus* evolved from *Australopithecus afarensis* in Model 1. OR *Homo erectus* evolved from *Homo ergaster* in Model 2. *or* Model 2 shows linear evolution from *Australopithecus afarensis*.

> For this part, any correct statement about the relationships illustrated was awarded a mark. These parts of the question did not require interpretation.

c Compared to *Australopithecus afarensis*, Homo erectus had *(two of)* (1 mark each):

- a larger brain case
- a less prominent brow ridge
- a more parabolic jaw.

Correct statements with respect to teeth size or position of foramen magnum received one mark for each correct feature named. Many answers were not structural, such as type of diet. Other incorrect answers included placement of eyes, sagittal crest and hair. Size of brain is incorrect as this can only be inferred; the size of the brain case is the feature.

d The arrow could be placed in either Model 1 or Model 2.

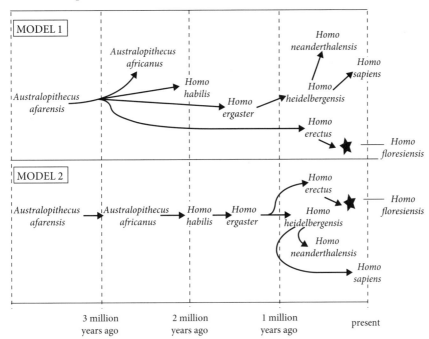

e **i** If several adult and child fossils of *Homo floresiensis* were found that had skulls indicating all had the same small brain characteristic.

  ii If other fossils were found in the same area and had normal sized skulls. If adult and child fossils of *Homo sapiens* were found in the same area and had skulls indicating child brain size much larger than that of the fossil called *Homo floresiensis*.

CHAPTER 5, UNIT 4 AREA OF STUDY 3

Multiple-choice questions

1 A

B is incorrect because it describes accuracy. **C** is incorrect because it describes repeatability. **D** is incorrect because it describes validity.

2 D

A is incorrect because it describes precision. **B** is incorrect because it describes accuracy. **C** is incorrect because this describes repeatability.

3 B

The tube in the light will undergo photosynthesis the whole time, decreasing the levels of carbon dioxide in the water.

A is incorrect because photosynthesis will stop at night without light. **C** is incorrect because the rate of photosynthesis will be higher. **D** is incorrect because the tubes in the darkness do not undertake photosynthesis.

4 B

Overnight, as a result of cellular respiration, the CO_2 produced turned the water yellow. As the day started and light increased, the CO_2 in the water was used up during photosynthesis and so the pH level increased, turning the water green by 9 a.m. If the darkness had been maintained, the pH of the water would have remained yellow.

A is incorrect because the chlorophyll will remain in the plant cells. **C** is incorrect because the rate of respiration remains constant and the pH will decrease. **D** is incorrect because the rate of respiration is higher than the rate of photosynthesis at night.

5 C

More trials are needed to improve repeatability and accuracy.

A, **B** and **D** are incorrect because they will not improve the results.

6 C

A is incorrect because the pH of the water was not measured. **B** is incorrect because the accuracy is questionable; the pH was not measured and amount of carbon dioxide dissolved in the water is unknown. **D** is incorrect because the experiment is repeatable, but there are a number of improvements that could be made.

7 C

Group 3's measurements are very similar to one another, so they are precise, but they are not the same as the electronic thermometer reading, meaning they are not accurate.

A is incorrect because Group 1's results are neither accurate nor precise. **B** is incorrect because Group 2's results are precise but not accurate. **D** is incorrect because Group 4's results are neither accurate nor precise.

8 D

Groups 4's results are most consistent between the three trials at each concentration, so are more reliable than the other groups' results.

A is incorrect because the results are valid but Group 2's results are less reliable. **B** is incorrect because Group 2's results are neither precise nor accurate. **C** is incorrect because Group 3's results are not as reliable as Group 4's.

9 A

The independent variable is the variable that the experimenter changes within the experiment.

B, **C** and **D** are incorrect definitions.

10 D

11 B

This is one way to investigate a mode of transmission.

A is incorrect because it is unethical. **C** is incorrect because pathogens are often host-specific so this human pathogen may not infect rabbits. **D** is incorrect because identifying the gene sequence is done through DNA sequencing, not PCR alone.

12 C

The presence or absence of antibodies is a qualitative measure. It is non-numerical.

A, **B** and **D** are incorrect as they are quantitative measures.

13 D

The results show that, with an increase in temperature, more of the agar plate is covered in bacteria.

A is incorrect because the percentages show the agar plates are not completely covered in the results. **B** is incorrect because bacteria do not need light to grow. **C** is incorrect because binary fission is not an exergonic process.

14 D

D is the variable that is measured, therefore is the dependant variable.

A and **C** are incorrect because time and number of bacteria are controlled variables. **B** is incorrect because temperature is the independent variable.

15 A

Students should repeat experiments to improve reliability.

B is incorrect because this is analysis of data. **C** is incorrect because this is an extension on the original experiment. **D** is incorrect because it is of no consequence.

SOLUTIONS

Short-answer questions

16 a Tube F is the control condition, used to make sure that the effects on the egg were not because of the water or other controlled variables. This tube is the one that all other experiments are compared against. The control demonstrates that the change in the dependent variable is due to the independent variable in an experiment.

b *Any three of* (1 mark each)*:* temperature, fluid volume, egg cube size/mass/surface area, light intensity

c No, the results do not support the hypothesis. (1 mark) The egg protein has been hydrolysed in tube B, which is the tube that contains the enzyme *and* the acid, resulting in the enzyme being in its optimum pH range. In other tubes that contain pepsin without the presence of acid, the egg has not been hydrolysed. (1 mark)

d This experiment is looking at the optimum pH range for the enzyme pepsin. (1 mark) Enzyme and substances that vary pH are added to each test tube to adjust the pH of the solution. (1 mark) The results show no egg in tube B because the enzyme is in it's optimum pH range with the addition of HCl, and as such the egg protein has been completely hydrolysed. This makes sense because the acidity of the stomach (pH 2-ish) is the optimum pH range for pepsin that breaks down proteins. (1 mark)

Note that each test tube in this experiment tells students something different:

Tube A – just enzyme at a neutral pH, therefore enzyme alone will not cause the digestion of protein

Tube C – just acid – enzyme is required to break down the protein, not just acid

Tube D – enzyme in alkaline pH is way out of its optimum range

Tube E – just alkaline environment

Tube F – control.

e The method should include (*for example*) (1 mark each) volumes of each component in each solution, size of egg white cube, resting temperature, or the inclusion of multiple trials; for example, three tubes for each condition.

f *Example answer:*

1 Set up the test tubes as follows: *(Note that all tubes contain 10 mL, and percentages tell the rest. For example, Tube A: 1.0% pepsin is 100 μL of pepsin in 9.9 mL of water; Tube B: contains 100 μL of pepsin, 40 μL HCl and the rest is water)*

A: 10 mL 1.0% pepsin

B: 1.0% pepsin in 0.4% hydrochloric acid

C: 10 mL 0.4% hydrochloric acid

D: 1.0% pepsin in 0.5% sodium bicarbonate

E: 10 mL 0.5% sodium bicarbonate

F: 10 mL distilled water (1 mark)

2 Cut six 0.5 mm cubes of cooked egg white and place one in each test tube. (1 mark)

3 Incubate at room temperature for 12 hours. (1 mark)

17 a It is hypothesised that if *H. pylori* is present in the stomach, it will cause stomach/duodenal ulcers.

b The experiment is valid because it does test what was intended (1 mark), but it is not reproducible because it is unethical to intentionally give humans disease. (1 mark)

c While this experiment clearly shows the relationship between *H. pylori* and stomach ulcers (1 mark), it is ill-advised to experiment on yourself because unknown consequences could have occurred. However, Marshall did have antibiotics to treat the *H. pylori*. (1 mark)

d Independent variable – presence of *H. pylori* (1 mark)

Dependent variable – presence of stomach ulcers (1 mark)

e   The data presented indicates there is a correlation between the presence of *H. pylori* and the development of stomach/duodenal ulcers. (1 mark) However, there is no mention of taking biopsies from healthy people as a reference for the presence of *H. pylori* in uninfected people. (1 mark)

f   In conclusion, there is a very strong link between the presence of *H. pylori* and the development of stomach/ duodenal ulcers. (1 mark) This supports the hypothesis that, if *H. pylori* is present in the stomach, it will cause stomach/duodenal ulcers. (1 mark)

g   It is not beneficial or cost effective to create a vaccine because *H. pylori* is not highly contagious, and it does not affect a large proportion of the population. (1 mark) It is cheaper to treat it with antibiotics rather than spend millions on the development of a vaccine. (1 mark)

18 a   *Either:*

- aerobic or cellular respiration
- anaerobic and aerobic.

b   Dependent variables: CO_2 and O_2 levels (1 mark)

Independent variable: temperature of the chamber/environmental temperature (1 mark)

'Heat lamp' and 'temperature' were not awarded any marks.

c   To establish a baseline for the experiment

d   *For example* (1 mark for the control measure and 1 mark for the explanation):

- Make sure that the experiment is conducted at the same time of day. Cockroaches have different levels of activity at different times of day, and this could affect rate of aerobic respiration.
- Feed the cockroach the same food each day to ensure the same initial glucose levels, which could affect cellular respiration.
- Ensure that the environment the cockroach is kept in between experiments is the same so that other factors such as external temperature do not affect the cellular respiration rate before the experiment.

e   i   When the temperature is constant, the levels of CO_2 increase sharply and slowly rise with decreasing temperature, (1 mark) and the levels of O_2 decrease sharply and then slowly decrease with decreasing temperature. (1 mark)

ii   Conclusion: Aerobic respiration is occurring, or the rate of cellular respiration is dependent on the temperature of the chamber. (1 mark)

Evidence: Oxygen is an input and is therefore decreasing. (1 mark) Carbon dioxide is an output and is therefore increasing. (1 mark) Low temperatures lower the rate of reaction. (1 mark)

This question gave students the opportunity to put the information together, drawing on the data given and then providing explanations and conclusions.

19 a   It is hypothesised that larger beans can be grown from the plants that produce the largest seeds.

b

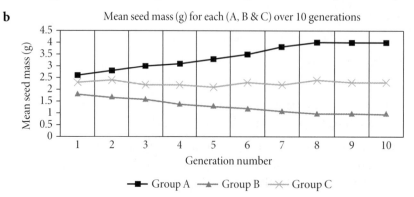

(1 mark for the *x*-axis, 1 mark for the *y*-axis, 1 mark for correct plotting of points)

c Group A seed size increased over multiple generations until they reached 4 grams and no further increase was obtained after generation 8. (1 mark) Group B results showed a decrease in seed size over multiple generations until a mass of 1 gram was reached at generation 8 and no further reductions were seen. (1 mark) Group C results showed very little variation in the size of the seeds over 10 generations. (1 mark)

> This question is asking for a description of results. When describing results in a scientific report, there is no evaluation or analysis included, simply describe the relationship seen in the results.

d Group C is the control group and is used as a comparison for Groups A and B. It is the 'normal' situation if there was no human interference.

e *Any three of* (1 mark each): soil type, nutrition provided to the plants, amount of water, day/night cycle or amount of light provided, temperature, size of the planting pots

f Environmental conditions can influence plant growth. (1 mark) If a plant is deprived of nutrients or sunlight, it will not grow as efficiently as one that has had ample of both. (1 mark) Therefore, the environmental conditions must be identical, so they do not influence the size of the plant and the size of the seeds produced. (1 mark)

g **i** The phenotype for larger seeds is influenced by many genes. The seeds grew larger because more of the genes for larger seed size were selected for and passed on to future generations.

  ii There is a balance between containing all of the requirements for germination and seed size. (1 mark) If the seed becomes too large then the germination process can be hindered due to the amount of energy required, *or* larger seeds can become a burden to the plant and influence the growth and health of the plant as they may be too heavy or require too much energy to produce. (1 mark) Therefore, the seed size stopped increasing after generation 8.

h Artificial selection

i Because this program looks at seed size, the plants produced may not be able to create very many seeds due to the energy demands. Therefore, productivity would decrease, or the health of the plant may deteriorate.

20 a It is hypothesised that the higher the concentration of antibiotic, the less bacterial growth will be observed on the agar plate.

b The independent variable is the concentration of the antibiotic. (1 mark) This is the variable that is changed during each trial. (1 mark)

c *Two of* (2 marks each: 1 mark for error; 1 mark for experimental design improvement):

- The bacteria were left at room temperature overnight – bacteria like *E. coli* need to be incubated at around 37°C for it to grow fast enough in large enough numbers overnight.
- She needed to let the bacterial broth dry on the agar plates before adding the antibiotic – otherwise, there will be a dilution in the concentrations added.
- She needed to allow the antibiotic to dry, otherwise the liquid would drip onto the lid when inverted, or spread across the agar plate.
- She needed to turn the plates upside down before incubating to lessen contamination risks from airborne particles landing on them and to prevent the accumulation of water condensation that could disturb or compromise the concentration of the antibiotic and the growth of the bacteria.
- She needed to use a different pipette for each of the different concentrations of antibiotic so that she did not affect the concentrations delivered to each plate.
- She needed to dispose of the gloves and then wash her hands (not the other way around) so as to make sure her hands were clean before leaving the laboratory.
- She needed to put on her gloves before she collected the bacterial broth for safety.
- She needed to repeat steps 5–8, not just 6–8, otherwise she may have omitted adding the bacterial broth to the plate.

d There would be no effect on the bacteria – there would be normal growth on the agar plates. (1 mark) Antiviral drugs are not effective against bacteria. (1 mark)